DIGITAL
BROADCASTING

Bloomsbury *New Media* Series

ISSN 1753-724X

Edited by Leslie Haddon, Department of Media and Communications, London School of Economics and Political Sciences, and Nicola Green, Department of Sociology, University of Surrey.

The series aims to provide students with historically grounded and theoretically informed studies of significant aspects of new media. The volumes take a broad approach to the subject, assessing how technologies and issues related to them are located in their social, cultural, political and economic contexts.

Titles in this series include:

Mobile Communications: An Introduction to New Media

The Internet: An Introduction to New Media

Games and Gaming: An Introduction to New Media

Digital Broadcasting: An Introduction to New Media

Digital Arts: An Introduction to New Media

DIGITAL BROADCASTING

An Introduction to New Media

Jo Pierson and Joke Bauwens

Bloomsbury Academic
An imprint of Bloomsbury Publishing Inc

BLOOMSBURY
NEW YORK • LONDON • NEW DELHI • SYDNEY

Bloomsbury Academic

An imprint of Bloomsbury Publishing Inc

1385 Broadway	50 Bedford Square
New York	London
NY 10018	WC1B 3DP
USA	UK

www.bloomsbury.com

First published 2015

Library of Congress Cataloging-in-Publication Data

Pierson, Jo.

Digital broadcasting : an introduction to new media / Jo Pierson, Joke Bauwens.

pages cm. -- (Berg new media series) Includes bibliographical references and index.

ISBN 978-1-84788-740-5 (paperback) -- ISBN 978-1-84788-741-2 (hb) 1. Broadcasting--History. 2. Digital television. 3. Digital audio broadcasting. 4. Television broadcasting. I. Bauwens, Joke. II. Title.

HE8689.4.P54 2015

384.54--dc23

2014042381

ISBN: HB: 978-1-8478-8741-2

PB: 978-1-8478-8740-5

ePub: 978-1-4725-1727-2

ePDF: 978-1-4725-1726-5

Series: Bloomsbury *New Media* Series

Typeset by Fakenham Prepress Solutions, Fakenham, Norfolk NR21 8NN

Dedicated to Lieke Pletsers

CONTENTS

ACKNOWLEDGEMENTS

First and foremost we would like to thank the series editors, Leslie Haddon and Nicola Green, for their excellent advice, unconditional support and enduring patience. We are also very grateful to the great team at Berg and Bloomsbury – especially Katie Gallof, Mary Al-Sayed, Laura Murray, Tristan Palmer, Claire Cooper and Kim Storry – for their feedback, support and patience.

We would also like to thank all our colleagues, in particular Wendy Van den Broeck, Bram Lievens, Iris Jennes, Karen Donders, Tom Evens, Nils Walravens, Pieter Ballon and Rob Heyman, for their invaluable inspiration and input. Our acknowledgement also goes to the trainees Tamar Betsalel and Tine Castro who have helped us in the preparatory data collection. We also wish to thank the Vrije Universiteit Brussel and the research centre iMinds-SMIT for being such a generous professional home.

Finally we would like to dedicate this book to our dearest families and friends, for enabling us to take the time and space to write this book.

1 INTRODUCTION

Broadcasting in the form of radio and television has evolved significantly since it was established during the beginning and middle of the twentieth century. Both forms of broadcasting are now being reinterpreted through the far-reaching effects of digitization and convergence. We are at a turning point with many promises and expectations, but where traditional broadcasting is still central in an economic, regulatory, social and cultural sense. Indeed, the oversimplified view that the current digital broadcasting evolution is marking the obsolescence of radio and television, typically voiced in industry analyses and market research reports, but also in academic research, is counterbalanced by other studies. This is especially true if we enlarge our view beyond the Anglo-Saxon part of the world and include economies, industries, societies and audiences (or consumer markets) that deviate from the trendsetting, early adopting and tech-savvy parts of the world.

Hence this book starts from the observation that digitization is producing some fundamental changes in the way broadcasting is organized, produced, distributed, received and consumed, but that technological change is always interacting with wider political, economic, social and cultural contexts, that help us put into perspective the dawn of a new age. Building on empirical studies and theoretical reflections on the phenomenon of broadcasting, most of them dealing with television, this book aims to encourage readers to reflect on how the transition from traditional to digital broadcasting is not only reconfiguring but sometimes also reproducing established arrangements of regulation and policy, industries and economies, production and content and audience practices.

The original and traditional notion of broadcasting dates back to modernity's restless and vehement escalation between the two world wars and after World War II, and is deep-rooted both in the anxiety of the then power elites for control over the populations, and the more noble democratic pursuit of a public sphere and spirit of community (Gripsrud 2010c; Scannell 1989; Williams [1974] 2003). Hence, the centre-periphery model of broadcasting is built upon the principle that symbolic content (i.e. news, music, talk shows, quizzes, soap operas) should be disseminated from one centre to many people, citizens, populations and consumers so that they would be informed, entertained and brought together in collective, simultaneous moments of reception as participants in a flow of continuous, sequential dissemination. Hence, modern citizens living their anonymous lives in large-scale, bureaucratized, complex, scattered and industrialized environments would still feel connected to a larger community of shared public interest (Gripsrud 2010c: 9).

Obviously, various developments are increasingly challenging this blueprint of broadcasting, not the least digitization that reverses or erodes the very foundations of broadcasting. With the World Wide Web and social media in particular one-to-many is converted into many-to-many; with the various new delivery platforms flows are interrupted and simultaneity is turning into on-demand; with the increase in the number of services, 'broad' is becoming 'narrower'. Is, then, digital broadcasting not a contradiction in terms? In this book we will argue that it is not, but rather, from a dialectical perspective, we will explore how, with the digitization of broadcasting, change and stasis – both in technological and social terms – presuppose and need each other. Hence the old is in the new and vice versa (cf. Van Den Eede 2012: 371).

This is clearly shown in the significant role television and radio are still occupying in everyday life. As broadcast media they remain easily accessible to all, in the sense of requiring fewer digital competences than computer technology. Meanwhile, television and radio have structured and – for the time being – still structure the everyday life, leisure time and the cultural habits of many people. Studies show that watching television (together with listening to the radio) at home is still the major free-time activity in many countries all over the world. Television as a media technology provides a so-called 'ontological security' for people, a secure knowledge that it is always there, precisely because of the extent to which it is familiar, indeed 'domesticated' (Silverstone 1994a; Silverstone 1994b). Equally, although people spend less time listening to traditional broadcast radio on broadcast bands in the radio spectrum (FM, AM and other bands), radio as a medium remains significant with a broad reach and substantial listening time. Thus it has become easier than ever for people to access music and niche content through the internet, while online radio use is increasing (Ofcom 2010).

Readers will be invited at many junctures to think about television and radio in other formats, and in different places, than they would 'traditionally' think about these media – making reference to technologies such as tablet computers, streaming media, Blu-ray discs, mobile television, MP3 players, smartphones, internet television and other emerging 'broadcast' media. Likewise, the pre-digital era of broadcasting will sometimes also be discussed in this book as a point of comparison, and in order to demonstrate that the transformations we are witnessing today were already emerging more than 20 years ago, with the various convergences and divergences of different platforms, technologies and software.

Digital transitions and other even earlier technological developments have already led to changes in the forms, functions and possibilities – or so-called 'affordances'

(Gibson 1977) – offered by broadcast media. Since the spread of video cameras and camcorders, audiovisual technologies have become more accessible for the larger public (together with user-friendly editing tools), extending the amateur filming community (making home movies, using low-cost toy video cameras and smartphones). Reception equipment has evolved from one central radio and television set in the living room, to multiple audiovisual receivers or screens throughout the house (e.g. in the bedroom, home office, kitchen). More and more households, at least in the more affluent parts of the world, have specific equipment at their disposal that enhances the multi-sensory and immersive experience of television and radio reception. Home surround-sound and digital music dock systems equal the sound and image quality of movie theatres and music studios.

Unlike these changes that point at a growing importance of so-called quality of reception experience, we also see that broadcast content like footage, music, clips and movies are produced, shared and consumed on different kinds of social media platforms (such as YouTube and Facebook). These are consumed as snacks, 'packaged bite-size nuggets made to be munched easily with increased frequency and maximum speed' (Miller 2007). The quality of the sound and image of this audiovisual content is less important than its lightness, brightness, digestibility, ease-to-access and novelty (see, for example, Shao 2009). More futurist developments point at converged multimedia environments in the home and – in the end – an internet-of-things smart space where data driven media channels, broadcasting channels included, will potentially act through algorithms as pervasive central gateways for communication and interaction with people and objects. We are already seeing how digital video recorders (DVR) connected to television sets (e.g. TiVo, Apple TV, network operator devices) and television via internet (e.g. Netflix, Amazon Instant Video) can recommend certain content and (re)organize the way programmes are watched, based on user data.

BROADCASTING IN TRANSITION

Drawing upon an interdisciplinary and international field of research and theory that deals with the impact of digitization on broadcasting and on television in particular, we take the transition to a digital era as our point of departure and pose the overarching question of what this means for various stakeholders involved: service providers, broadcasting companies, infrastructure and platform providers, media professionals, policymakers, researchers and, last but not least, large groups of people all over the world who are watching and making use of television and radio. The transitions are often described as major, innovative and disruptive – and to a certain extent rightly so because the relatively long domination of analogue radio and television (more than 70 and 50 years respectively) is coming to its end. Hence various scholars, policymakers, industry spokespeople and media have been announcing for more than ten years now either the death or the reinvigoration of broadcasting as an economic system, cultural form and social institution (Katz 2009).

However, as with many other media evolutions, change and novelty are far less revolutionary than often believed, or at least happen at a far slower pace than heralded. Let us, for example, consider the slow adoption of digital television technology among viewer audiences (Van den Broeck and Pierson 2008). In the US, the first digital satellite television service was offered in 1994, and in 2000 digital High Definition television was introduced. These digital television technologies enhanced the quality of reception, and of image and sound; brought about an increase in the number of available television and radio channels; facilitated access to better information services; and enabled interactivity (e.g. on-demand services and polling-voting) (Sourbati 2004). Yet, European audiences were not immediately willing to invest money in the new technologies, as they did not see an added value that justified the costs (Iosifidis 2006). Obviously, the existing broadcasting system and service also play a crucial role in the pace of digital broadcasting adoption. In countries that have a paid television tradition, digital television is more quickly embraced than in countries where viewers are used to free-to-air television. In 2007, IP Network, the leading European broadcast advertising network owned by RTL group, estimated that the average penetration of digital television in the 26 EU countries (Malta not included) was 40.8 per cent (IP Network 2008). Figures from 2013 by the research and consulting institute IDATE show that 65.2 per cent of the world's households have access to digital television. They are expected to rise to 92 per cent in 2018 (Ollivier and Pouillot 2014: 62). It was only by packaging different audiovisual, internet and telecommunication services as one product and in this way offering one or more of these services at a decreased price – often marketed as 'triple play' (combining broadband internet access, television and telephone) or 'quadruple play' (when mobile phone service provisions are added) – that digital television was adopted more rapidly. Hence, we will see throughout this book that traditional radio and television are still not dead, and that the 'core idea' (cf. Peters 2009) or socially accepted use of these media is not easily replaced by new meanings and social uses.

At the same time, broadcasting media, which used to be so familiar and obvious to us, are turning into new media, i.e. 'media we do not yet know how to talk about' (Peters 2009: 18). In particular, technological, political, economic, social and cultural developments related to the internet and telecommunications are strongly affecting the media we used to know so well. Mackay and O'Sullivan (1999: 4–5) describe digital television as an 'old' medium in 'new times', a phrase that captures well what television as a social institution, as an industry, as a technology and as a social-cultural practice is going through in 'the overall, largely digitized media system where the internet now plays an important role' (Gripsrud 2010b: xv). In other words: we are witnessing the 'refashioning' and 'remediation' of a medium that tries to answer the challenges of new media (Lister et al. 2003: 39-40).

For example, one of the shifts technology developers and cultural industries are interested in is the proliferation of different networked devices in the living room, and how this might affect the modes of watching and using television. More and more people already use their computer, smartphone or tablet as a second screen while watching television, and many developers are exploiting this in order to create applications that enable interaction related to television content. Moreover, interactive digital broadcasting is moving beyond the set-top box, as connected television sets and applications like internet television (e.g. Google TV, Apple TV) link traditional television with the internet. In industry milieus it is often believed that these developments, which show a move towards device divergence instead of device convergence, will change the way people experience the role of broadcasting. In this respect data driven personalized services and recommendations systems will need to take into account the multiple users that are often present when watching television. On the other hand, connected devices enable social interaction at a distance, allowing communication and other remote interactions to take place through or alongside television.

As such, digital broadcasting might be a good example of 'media renewability' (Peters 2009), a process that passes through five periods: (1) technical invention; (2) cultural innovation; (3) legal regulation; (4) economic distribution; and (5) social mainstream (op. cit.: 18). As will become clear, the writing of this book and the knowledge we are building upon is situated in the transition between the first period – 'during which media are recognized rarely as "new" and usually thought of as "old plus"' (op. cit.: 18) – and the next three stages, during which broadcasting is developing 'new social uses', and 'the interested parties explicitly contest and negotiate for media power', both in legal and economic terms (ibid.). The last period, 'social mainstream', has for the time being still not arrived in large parts of the world. Yet small groups in affluent societies already consider digital broadcasting as no longer new.

Perhaps McLuhan's four laws of media, which he himself, medium determinist par excellence, has called the 'tetrad of the effects of technologies and artefacts' (McLuhan and McLuhan 1988: 99), provide a better theoretical framework to get a grip on the ambiguity and insecurity that media in transition create. Rather than seeing media evolutions and media effects as a sequential process, McLuhan sees a cluster of four processes simultaneously at work, which are all in the medium enclosed from its very start, i.e.: enhancement, obsolescence, retrieval and reversal. 'Usually', he continues, '"media study" (and equally promotion) covers only the first two aspects' (ibid.). That is, we are particularly interested in how a new medium enhances, intensifies, accelerates and makes things possible. Conversely, we also pay a lot of attention to the complementary process of enhancement and chiefly investigate what is displaced, pushed aside or obsolesced by the new medium. However, in McLuhan's theory, obsolescence is not the end, but the beginning of bringing old things up to date, retrieving them in a new form, modifying them so that they can be part of the new environment, culture or meaning (i.e. retrieval). The fourth aspect of the tetrad comes closest to what, in marketing terms, is often labelled as revolution. Reversal is what happens when a medium is 'pushed to the limits of its potential, reverses its characteristics and becomes a complementary form' (McLuhan and McLuhan 1988: 107).

Introducing McLuhan, who has been often been described as a prescient thinker (cf. Silverstone 1988: 389), does not imply that this book's approach will be only forward-looking, although it covers the potential and forthcoming economic, social and cultural shifts in broadcasting. Rather it gives a commonsensical and evidence-based overview of how the transition from analogue to digital broadcasting is coming about in the broader context of industry developments, policy choices and social-cultural developments, as well as how it is affecting the processes of production, distribution, reception and use. Building upon the above-mentioned theoretical insights on media evolutions, we will demonstrate that radio and television have been dynamically renewing themselves over time, perhaps less radically than today, and that the novelties and changes they have gone through have been the precursor of the transitions we are witnessing today. We find that although change is never instant, it is nevertheless from time to time rapid; that structures sometimes impede change, sometimes speed it up; that television is still television to many people, and at the same time breaks free from its traditional meaning. In dealing with all these juxtapositions the book aims to give an informed account of how social institutions – i.e. citizens, consumers, policymakers, civil society, industrialists and engineers – are in a state of flux while dealing with these broadcasting changes.

A MULTIFACETED APPROACH TO DIGITAL BROADCASTING

By examining and integrating the different processes that accompany the digitization of television, we aim to contribute to a multifaceted understanding of 'the conditions and forces that cause the changes and orient their evolution' (Buonanno 2008: 59). One constant explanation in the history of all new media is unquestionably the interplay of economic and political powers as the impetus for media evolutions. The digitization of television definitely leads to new economic and innovative opportunities for the audiovisual and telecom industries. This explains why in some countries these industries, and in particular satellite companies, have initiated digital broadcasting. One of the first countries in which digital television became successful was the UK, where digital satellite television had already been introduced by 1988.

In other European countries (e.g. Sweden, Denmark, Greece and Spain) the first initiatives didn't appear until the end of the 1990s, but here satellite companies were also initiating the digital services, albeit not always with the same success in terms of subscriber numbers. However, when it comes to the usage of digital video broadcasting in European households, satellite is the most widespread technology, as 42 per cent of digital homes watched digital television via satellite, 38 per cent via terrestrial TV, 16 per cent via cable and 4 per cent via IPTV (IP Network 2008). Global figures from 2012 show that 93 per cent of the satellite network is digital. This is in sharp contrast with the other systems, as 50.5 per cent of the cable households and 65 per cent of the terrestrial households still have an analogue service (Ollivier and Pouillot 2013: 52).

Clearly, transitions in the policy landscape are closely linked with changes in the television industry. The audiovisual broadcasting industry and infrastructure have long been separated from other ICT industries like computers and telecommunications. However, owing to socio-economic, political and technological processes, these industries are gradually merging. We notice how the digital and converged television technologies shape and are being shaped by government policies and industry strategies (Flichy and Libbrecht 1995). The mutual shaping processes taking place between government policies and digital broadcasting development can be seen clearly in events such as the policies on the digital 'switchover', defined as 'the progressive migration of households from analogue-only reception to digital reception' (BIPE 2002). But the relationship between policy and industry is also obvious in the licensing of the UMTS (Universal Mobile Telecommunications Service) spectrum (via auctions looking for the highest bidder, via 'beauty contests' assessing the most attractive offer, etc.) and the 'digital dividend' (i.e. digitization takes less space in the broadcasting spectrum than analogue, which creates a surplus or dividend for other digital services). Government and regulatory decisions related to the digital dividend are opening up certain pathways, but closing down other ones. For example, freeing-up the airwaves previously occupied by analogue television (following the digital 'switch-off') creates space for digital television services (like mobile television), but can also create opportunities for other services such as the wireless internet (e.g. Google WiFi bids in the US). At the same time we see how technological changes within industry and in the infrastructure have had an influence on public policy and the regulation of broadcasting. This refers, for example, to the public service remit in a digital and converged media environment in comparison to the analogue monopoly situation of the past. As public service broadcasters venture into digital services (e.g. online news offerings, developing software applications (apps), etc.), these services need to have a public value in order to justify the public money for these extra activities. Hence, in the UK, Germany and Norway, for example, so-called public value tests have been developed at the insistence of the

European Commission in order to assess and control the expansion of public service broadcasters into new audiovisual, mostly digital, services (Moe 2010). The rationale behind this is the Commission's concern about the unfair competition public service broadcasters might cause to their commercial pendants through the additions of, for instance, press- or game-like services (Donders and Pauwels 2008).

Although industry and policy play an important part in media evolutions, the social-cultural conditions of everyday life, routines, rhythms and rituals, social inequality and cultural diversity also help us understand why some media uses are remarkably persistent, and others far more easily pushed aside, retrieved or reversed. Since the 1950s, television – in particular – has established itself as a daily, domestic, routine and structuring leisure activity. Families have gathered around the electronic hearth to spend their evenings at home (Lull 1990). Billions of citizens have participated in mass ceremonies on their couches, becoming part of national and international communities (Scannell 1989). For many people, television programmes have provided 'food' for conversation among family members, friends, relatives, colleagues, neighbours and passers-by in shops, trains and buses (see among many others: Gillespie 1995; Hobson 1989; Paterson, Petrie, and Willis 1995). Studies and commentaries on the meaning of television in people's everyday life often question if and how digital television will affect the medium's role in family life as a sort of 'intergenerational glue' between parents and children, and in community life as a cross-demographic bridge binding different social groups (Hartley 1999). Although empirical evidence indicates that the dispersal of media tools and their use at home, as new screens (i.e. laptop, games consoles, tablets and other handheld devices) and television peripherals (DVD players) offer an enormous array of media consumption opportunities, it is still unclear how families themselves are dealing with the bonding potential of television within their daily routines. When it comes to national communities, the number of television channels available has proliferated significantly. A vast gamut of network and terrestrial television stations, independent satellite and cable channels, generalist and specialist television channels, public service and commercial broadcasting channels, and pay-per-view and on-demand delivery systems increasingly construct segmented audiences. As a result, the simultaneity of the television experience (watching the same programme at the same moment) crumbles in an age of channel abundance (Lister, Dovey, Giddings, Grant, and Kelly 2003: 30).

From a production perspective, the internet, more particularly Web 2.0, and the coming of easy-to-handle recording devices, have brought about a media culture in which the clear-cut distinction between media professionals (who make the media) and media amateurs (who consume the media on offer) has been called into question. Hence, traditional bastions of audiovisual production, such as broadcasters, studios, production companies and professions, are increasingly confronted with this new world of networked connectivity and 'co-creative potential [...] captured on the web' (Dovey and Rose 2012: 159). Surely, this has not led to the waning of professionalism or the disappearance of traditional, good old-fashioned approaches to audiovisual formats, genres and content. Soap operas are still being made and broadcast, music charts can even at this time still be heard on innumerable radio stations, and newscasts have not vanished from radio and TV. Rather, the fusion of these two spheres of popular media culture production, i.e. broadcasting and internet, is bringing about interesting, contradictory and unpredictable developments. Some scholars, with Jenkins (2006) as the most notable exponent of this strand of thought, notice the rise of a new paradigm of production and consumption, i.e. convergence culture, which is based on participation in all its meanings, ranging from the re-appropriation and re-working of media industry texts by enthusiastic, playful and critical fans (e.g. frivolous dubbing of film scenes; relocating popular culture characters in new contexts; suggesting alternative endings to blockbuster films as by HISHE.com <How It Should Have Ended>) to the making, uploading and sharing of original audiovisual material made by creative people with various backgrounds: not (yet) established professionals, artists (as for example the Swedish artist Anders Ramsell's Aquarelle Paraphrase of the motion picture *Blade Runner*), students and many media users involved in 'mini c-creativity' (see Beghetto and Kaufman 2007) or 'vernacular creativity' (Burgess 2006), all showing and sharing their creative work on YouTube. Other scholars see in the intersection between web culture and broadcasting industry a lot of signs that the latter is successfully recuperating the creative potential of people and using all their stimulating, enthusiastic and spirited work as a form of 'free labour', i.e. pleasurable media work that is not paid for (Terranova 2000), from which the big corporations can gain profit, steal ideas, market their brand image and secure their position (Caldwell 2006; Örnebring 2007a).

OUTLINE OF THE BOOK

In Chapter 2 we step back in time to look at the history of broadcasting. Chapter 3 then sets digital broadcasting within a broader industrial and infrastructural framework by describing the key technological changes taking place in distribution, transmission and reception, and relating this back to the policy and regulation presented in Chapter 2. Chapter 4 discusses transitions in the production process and the resulting output (like television formats and popular genres) in relation to the evolution of digital broadcasting. Chapter 5 deals with the question of how the meaning, position and role of broadcast channels is challenged and possibly redefined within the digital ecosystem. Chapter 6 discusses how the use of broadcasting media texts and technologies interrelates with the everyday life of individuals and households. Finally, we take a closer look at and critically assess future trends for broadcasting audiences. Chapter 7 concludes the book with a discussion of how all these digital transitions have shaped theory and research. Fundamentally, the issues discussed question how the scope and nature of traditional television and media studies have to change. First, however, we turn to discuss and contextualize the social history of broadcasting, and provide a longer-term understanding of changes in broadcasting, both as a system and as a technology.

Chapter Summary

■ The aim of this book is to understand how broadcasting in the form of television and – to a lesser extent – radio are currently being reinterpreted through the far-reaching effects of digitization and convergence. We argue that 'digital broadcasting' is not a contradiction in terms, but – on the contrary – both terms presuppose and need each other.

■ Drawing upon an interdisciplinary and international field of research and theory, the aim of the book is to give an evidence-based overview of how the transition from analogue to digital broadcasting is coming about in the broader context of industry developments, policy choices and socio-cultural developments, as well as how it is affecting the processes of production, distribution, reception and use. We are witnessing the 'refashioning' and 'remediation' of a medium that tries to answer the challenges of new media.

■ This book first looks at the history of broadcasting in order to better understand the origin of the current broadcasting landscape (Chapter 2). The next chapter (Chapter 3) then takes an industry and infrastructural perspective by focusing on the main technological changes in distribution, transmission and reception. The latter is related back to the policy and regulation presented in Chapter 2. In Chapter 4 we discuss changes in the production process and the resulting output, linked to the evolution of digital broadcasting. This is followed by the question of how the meaning, position and role of broadcast channels is challenged and possibly redefined (Chapter 5). The audiences and how their engagement with television and radio is changing (or not) is the main topic of Chapter 6. Here we investigate how the use of broadcasting media texts and technologies interrelates with everyday life. Finally, in Chapter 7, we examine and critically assess future trends for broadcasting audiences, and their possible implications for theory and research.

2 A HISTORICAL APPROACH TO DIGITAL BROADCASTING

Today's use, place and meaning of traditional broadcasting media in people's lives, shows that radio and television have become deeply domesticated technologies, in terms of ubiquity, familiarity, everydayness and ordinariness. Obviously, both media have always had a strong connection with the domestic sphere, although, as will become clear throughout this chapter, this was not inscribed in the 'nature' of broadcasting and the consumption of radio quickly expanded to other spheres outside the home. We will therefore use the concept of domestication here in its extended definition (Haddon 2003), and argue, following in others' footsteps (Moores 1993; Spigel 1992), that the entanglement of broadcasting and the domestic family life is a crucial component of broadcasting's 'core idea' (see Chapter 1). In order to understand both the renewability and resilience of broadcasting, it is important to think through the dialectical relationship between the private domain and public domains of everyday life, and the reconfigurations of both domains. The historical domestication of broadcasting has entailed a number of processes. First, it involved the interplay between the daily rhythms of life inside the home *and* outside the home. Second, it included the interaction between the social textures of relationships within the household (family, un/married couple) *and* beyond the household (friends, colleagues, fellow-citizens, fellow-consumers). And, third, it encompassed the context of the collective environment in which new radio and TV technologies have been repeatedly installed and appropriated.

This chapter gives a helicopter view of the historical development of broadcasting as a social institution or, as Williams ([1974] 2003) would have preferred, a 'cultural form'. Consequently, the focus will be on the complex process of incorporating radio and television into everyday life, on the larger social climate of cultural ideals (cf. Spigel 1992) in which the installation of new radio and TV developments time and again take place, and on the regulation and policy interactions with technological and economic developments. Being the most omnipresent media in the lives of large populations all over the world, this chapter explores how the old, established and accepted meanings, systems and regulations of broadcasting have been challenged and renewed throughout history.

THE INSTALLATION OF BROADCASTING MEDIA

Television and radio broadcasting were introduced into the home around six and nine decades ago respectively. A good account of the introduction of television as a domestic technology can be found in Williams' book *Television: Technology and Cultural Form* ([1974] 2003). Williams describes a complex of developments at the end of the 1920s, i.e. the coming of the motorcycle and motor car, the box camera and its successors, home electrical appliances and radio sets, later called 'consumer durables'. In a social sense this range of goods was characterized by two tendencies of modern urban living that look paradoxical, but are in point of fact intertwined: mobility and the inward-organized family home. Williams ([1974] 2003: 19) grasped this trend with the concept of 'mobile privatization', as it described a way of living that was both, in various senses, increasingly mobile and at the same time increasingly home-centred. Broadcasting was a social product of this paradox as it offered people a way to link an increased sense of privacy and search for safety, as manifested in the private family home (see also Spigel 1992), with the new large-scale organization of work and government in modern industrial societies. Hence, broadcasting, first via radio technology and later by television, served the social need for bringing the fast-moving and complex outside world into the privatized home (Gripsrud 2010c). Through offering a wide range of information and entertainment programmes, families were invited to gather around a new technological piece of furniture, like a new stove or fireplace, a focal point that helped them intertwine their personal and private lives with the larger structure of the nation-state recovering from the war and preparing for another: family togetherness as a tool for national community spirit (see also Boddy 2003; Douglas 1999; Gray 2003; Moores 1993).

As pointed out by others, there was nothing in the inherent technological properties of radio and television that determined it should be a medium of family togetherness (see Ellis 2000; Moores 1993). The history of radio indeed shows that the medium was originally a do-it-yourself-toolkit that radio amateurs-experimenters, enthusiasts and hobbyists, mainly men and boys, operated by transmitting and receiving signals via headphones. After World War I amateur experiments were if not banned then at least carefully watched by the government (Coe 1996; Street 2006), and radio became a real broadcasting medium (instead of a technical gadget or machine) (Regal 2005; Spigel 1992). With the formation of commercial radio stations and public service broadcasters and the expansion of programme types and schedules in the 1920s, radio itself was conceived of as being a full member of the family, with loudspeakers so that the whole family (women and children too) could listen to it (Moores 1993; Spigel 1992: 27; Taylor 2002). The history of television also shows that the first television sets in the mid-1950s – in countries like the US (Spigel 1992), Japan (Yoshimi 2010: 540–3) and Italy (Buonanno 2008: 14) – were not placed in private homes, but were to be found in outdoor places, from railway stations, parks, squares and even shrines and churches (public locations) to taverns and restaurants (semi-public places), where large numbers of people gathered to partake in TV exhibitions as if they were attending an open-air theatre. As soon as more households acquired a TV, the homes of these people became improvised mini-theatres, where relatives, family and neighbors were invited to watch this new magic, modern medium (Buonanno 2008: 15).

According to Williams, the domestic TV set that was developed for the consumer market was in fact an inefficient medium for visual broadcasting since its visual quality was poor and remained inferior to the quality of cinema despite later enhancements (e.g. colour, high definition) (Williams [1974] 2003). In spite of this, in the specific social context of the privatized home, television provided what Williams (op. cit.: 22) calls 'a wider social intake' by offering, in contrast to cinema, different types of content (e.g. music, news, entertainment, sports) and enabling people to engage with this in a private mode of media consumption, for which they already were socially prepared because of radio (cf. Taylor, 2002). Just as the diffusion of radio in Western societies was wrapped up in post-war discourses of modernity, the march of television too was accompanied by a strong belief in progress and reconstruction after World War II, resulting in a fast and cross-demographic spread of the medium. For example in 1953, the official launch year of Belgian public service television, only 6,500 households (0.1 per cent of the population) owned a TV set; by 1958, the year of Brussels World's Fair, the number of families with a TV had increased to 223,168 (2.5 per cent); between 1958 and 1960 the number of TV sets had multiplied by five (Bauwens 2007). Whereas only half a million of the Dutch population had a TV set in 1958, ten years later 80 per cent of households in the Netherlands were equipped with their own TV set (Righart 1995: 56–7). In the UK, only 0.3 per cent of the population had a TV set in 1948, but by 1958, this was already 52 per cent, rising to over 90 per cent by the 1970s (Hamill, 2003).

Reflection: Changes in TV and Radio

Think of yourself (or your parents, grandparents or other family). Can you:

a) Describe how the role of television in everyday life has changed over the years?
b) Describe how the way you listen to the radio and consume music has changed (or not)?

BROADCASTING AND CHANGING CONCEPTIONS OF THE HOME

Spigel's (2001) account of the introduction of television into the post-war US home is instructive, allowing us to comprehend how contemporary views and uses of digital broadcasting in everyday life are part of a wider genealogy of ideas about the nature of domesticity in a 'media-saturated world' (Couldry, 2012). In particular her cultural-historical approach to the so-called media home, i.e. the way domestic architecture and electronic communication are interlinked and give shape to housing and household types, is helpful to understand how people have put television and its peripherals to specific social uses, so that it has become omnipresent, fully objectified and incorporated in everyday life. Objectification refers to the way in which the object of television is fitted in and adapted to the household culture, at first by being appointed a specific, meaningful place in the domestic geography. Incorporation relates to functionally fitting in the use of television within the household member's rhythms, routines and time allocation (Silverstone and Haddon 1996: 45–6).

Following Spigel, the media home has historically been built upon three cultural metaphors through which domesticity and media have been imagined, that is: theatricality, mobility and sentience. These three cultural images have been central to the increasingly linked histories of domestic architecture and electronic communications, and they have materialized in three different media housing types, to be elaborated on below: the 'home theatre', the 'mobile home' and the 'smart home' of the digital future. Interestingly, we see how each of these metaphors, which exemplify the successive conceptualizations of television in the past, is coming back in the contemporary age of digital broadcasting. Hence, the 'smart home' links up with current discussions of trends in digitization, while the 'home theatre' and the 'mobile home' return in current debates on digital television and in the development of digital broadcasting tools.

Reflection: Different Perspectives of the Home and Digital Broadcasting

Before looking at what these metaphors refer to in the next section, can you anticipate:

a) In what sense can digital broadcasting be a manifestation of the notions 'home theatre', 'mobile home' and 'smart home'?

b) Can you think of other meanings that television and radio could have in a fully digital era?

In the 1950s, when TV was praised as a new entertainment and information medium, the two central ideas that guided the discussions about TV were its theatricality and mobility, which Spigel denotes as 'the home theatre'. This concept refers to the idea of TV bringing amusement into the living room and providing family members with a window on the world. In receiving information from outside the house, viewers were in their imagination 'transported' around the globe without leaving the safety of their home. In the 1920s radio broadcasting also played on this double potential of opening up the world without having to leave the house. In particular in the US, the social conditions of the 1920s (large rural areas, predominantly non-electric and media-poor farmer households with only newspapers and no movies) paved the way for a medium that 'opened up new vistas for isolated families' in search for recreation and connectedness in their spare time (Coe 1996: 27). As the cultural ideal of family life and domesticity grew in importance in the modern Western world and the colonial parts affected by Western industrialization, television filled out this conception of broadcasting as a 'unifying agent' of both family togetherness and 'democratic harmony through the mass dissemination of culture' (Spigel 1992: 3). Advertisements promoted TV as a form of 'going places', without having to spend money or effort to actually go there (Spigel 1992, 2001).

In the era of digital broadcasting, the 'home theatre' metaphor now refers to increasingly simulating the (movie) theatre experience by way of enhanced viewing experiences with high-quality images (e.g. HDTV, Ultra High Definition Television) and surround sound (e.g. Dolby Surround). Other innovations from cinemas, like 3D, can now also be brought into the home with 3D televisions. Through the digitization of images and sound, transported over networks, and delivered via middleware, set-top boxes and devices, the digital broadcasting industry, in particular device manufacturing companies, is spending a considerable amount of effort and money to enrich the so-called quality of experience.

The notion of 'homecasting', introduced by van Dijck (2007a, 2007b), is a good illustration of how the relationship between domestic space and broadcasting is being rethought in the age of digital broadcasting. By making use of video-sharing websites, pre-eminently YouTube, to 'download and upload prerecorded, rerecorded, tinkered, and self-produced audiovisual content via personal computers' (van Dijck 2007a: 4), the home itself has become a centre and channel that reaches out to 'anybody's home' (op. cit.). In some cases the home has turned into a potential firm where professional amateurs, the so-called pro-ams, or semi-professionals try or hope to make money with their home-made media content (Burgess and Green 2009a). Hence, from an exclusively reception-oriented context, the home has in part become a unit of production and distribution.

The appropriation of the television set in the home, and its accompanying symbolic representations and discourses in the media, advertising campaigns and everyday life, evolved together with changes in housing styles and ideals. At the end of the 1950s, the notion of home theatre made room for a new notion of the house, what Spigel (2001) defines a 'mobile home'. In advertisements and magazines this new housing style was depicted as a vehicle for transport and more concretely houses were designed with clear references to spaceships and rockets. The display of new technologies became important and specifically the newest object form of television, the portable receiver, was notable in this regard. Television sets evolved into mini-portables advertised through metaphors of transport supporting new cultural fantasies of travel away from home. The mini-portables could be taken outside transforming the idea of bringing public entertainment in the home towards ideas of taking the interior world outdoors. Spigel (2001) refers to this as the idea of 'privatized mobility', reversing the idea of William's 'mobile privatization', to express how people experienced the home as a mode of transport in itself that enabled them to take the private life outside. The merging of indoor-outdoor and the outdoor lifestyle became important in housing, while television sets were depicted in advertisements and magazines as perfect outdoor tools.

Obviously, this trend towards mobility was also shown in radio's development, and was more successful than TV's early experiments with portable television sets. Indeed research shows that the portable television sets, which were heavily marketed in the 1960s, were in practice seldom moved (Spigel 2001). By the early 1930s, radio enthusiasts had already adapted domestic radio equipment to use in cars. Commercialized in the mid-1930s, the car radio became a common car feature in the 1950s and 1960s, with push-button station selectors. Resilient and adaptable to the new cultural, social, economic and technological challenges (Matelski 1995: 5), radio became a 'truly ubiquitous mass medium' and 'personal service' when the transistor made its entrance in 1947 and become massively popular in the mid-1950s, especially among teenagers (Bull 2004; Crisell 1994). It 'had the effect of freeing the radio from the confines of the home' (Berry 2003: 283), and increased the phenomenon of what the industry called 'out-of-home' listening (Douglas 1999: 221). Although the tablet computer and smartphone are increasingly used for watching (online) television content as a portable television set, listening to the radio still seems to be much more mobile, flexible and non-domestic than watching TV, since research indicates that even today's small and handy portable television devices are in practice used more often in the home than outside it (Södergård 2003; Vangenck et al. 2008).

On the other hand, mobile radio consumption has been dramatically challenged by the arrival of other mobile sound technologies, from the Walkman right up to today's MP3 devices, of which the iPod is probably the most paradigmatic example of a new culture and experience of music consumption, driven by miniaturization, mobility, power capacity and personalized playlists (Bull 2006). For example research in the US shows that iPod use in the car is often seen as a means of escape from commercial radio broadcasting, with its cutting off of the beginning and ending of songs and interruption of programmes with advertising breaks. Families are also creating their family iPod-playlists ('something for everybody') for long car trips, which like radio in the interwar years also produce a sense of shared family experience (Bull 2006: 141).

Reflection: Current Meaning of Mobile Privatization and Privatized Mobility

Given the ways in which we can increasingly access digital content on portable devices in a variety of places (e.g. watching a TV series on your laptop when commuting between home and office; following a political debate on your tablet while waiting for the bus; running in the park while listening to your favourite podcasts), think about whether Williams' 'mobile privatization' and Spigel's 'privatized mobility' are still relevant for capturing the societal meaning of digital broadcasting.

A more extended definition of portability in relation to digital broadcasting can perhaps be found in current debates about the so-called 'second screen'. This refers to using a second device, most often a tablet computer (e.g. Apple iPad), while watching television. Second screens are used to interact more intensely with the programmes on television, like commenting, voting, connecting with others, using social media (e.g. Twitter), having access to background information and shopping (through interactive advertising for example) (see also Chapter 4). While in the past interaction with television content was done through postal mail and texting (e.g. SMS voting), multi-modal use of 'second screen' devices transports the viewer into other spheres related to the programme.

With the 'smart home', the third type of media home, the modern ideals of mobility, freedom and progress still remain important in contemporary visions of domesticity. Yet the developments in telecommunications offer more than just mobility; they also make possible the construction of sentient or 'intelligent' spaces. Here digitization (and automation) has become central, but the twin desires for domestic comfort and family harmony, on the one hand, and participation, travel and mobility in the world outside, on the other hand, remain important narratives in the stories society is building about media technologies (Spigel 2001).

Obviously, the notion of the 'smart home' strongly resonates in the digital broadcasting era and can be connected to older concepts of intelligent homes and domotics (Harper 2003), which entailed a vision of interconnected devices in the home, potentially centrally controlled within the home or remotely controlled from outside it (Cawson et al. 1995). In particular, the relatively more recent development of 'ubiquitous computing' (i.e. the disappearance of the infrastructure into the background) (Weiser 1991), ambient intelligence (Aarts and Marzano 2003) and the internet-of-things environment (Gershenfeld et al. 2004) typify the coming of the 'smart home'. Although this type of house is still 'depicted as a toy for the wealthy' (cf. Spigel 2001: 401), many household appliances and other objects in the house (e.g. refrigerators, heating systems, security systems, etc.) are getting 'smarter' by connecting them to the internet via sensors and actuators (e.g. RFID tags) and, in terms of costs, becoming more available to the average middle-class family. When it comes to TV, two particular developments link up with contemporary society's preoccupation with time-saving technologies that integrate as many functions and tasks as possible and make life much easier.

First, the so-called 'smart TV', also known as the connected or hybrid TV, integrates the internet into television sets (e.g. with Google Chromecast), set-top boxes and DVD-players and sends TV to other screen devices both inside and outside the home (Wii, Xbox, laptop, tablet, smartphone). It offers a – in marketing jargon – 'one-touch' access to apps like iPlayer, streaming video-on-demand services such as Netflix, Amazon Instant Video and Blinkbox, and of course YouTube. The scattering of TV across multiple screens and the expanding on-demand engagement with broadcasting content definitely reconfigures the traditional domestic role of broadcasting media and television in particular. Where television has often succeeded in bringing all members of the household together, research suggests that the younger media generations within families are increasingly making use of their personal media or the family's computer screen devices to watch television content individually (Livingstone 2007; Roberts and Foehr 2008) (see also Chapter 6).

Second, the technological convergence of ICT and TV opens the door for what we would call 'algorithmic TV'. By registering and knowing the viewing preferences of the household members it detects in front of the television set, the data driven broadcast content and advertising can be attuned and personalized to the characteristics, tastes, consumption styles and moods of the audience. Obviously, the on-demand and interactive engagement with broadcasting content on the internet, which offers more opportunities for consumer choice and satisfaction, also entails an irreversible surveillance and trade of personal data, also known as data-mining (see Chapters 3 and 5), making the domestic sphere of private media use more and more knowable to and potentially more manipulable by the industry.

MODELS OF BROADCASTING SYSTEMS

Like the technologies, the broadcasting systems too are the historical result of (sometimes very) different industrial, economic, cultural and policymaking processes. For a global sketch of the transformations within the digital broadcasting industry it is useful first to make a basic distinction between the two opposing models on which broadcasting systems worldwide have been and still are organized. The first is built on government funding and a form of government intervention, which has varied, and in some former Communist countries in fact evolved, from strictly state-controlled television and radio to public service broadcasting. Within the second model, broadcasting is conceived of as a commercial activity (commercial broadcasting, private stations), with the notable example of the US, where from the very start the broadcasting system was based on advertising revenues. Today, the two models are often mixed, with most of the public service broadcasters also relying on commercial revenues and with private radio and TV stations licensed and controlled by governments (as for example in the Russian Federation). Clearly, the two models, and their variants, are linked to political ideologies and regimes, where states hold different opinions and normative beliefs about how society and the market should be organized. Today, the public service and commercial broadcasting models are dominant. If we take respectively the UK and US as the canonical examples of these models, most other broadcasting systems in the democratic countries of the world are situated somewhere in-between.

The UK broadcasting industry has been characterized by government control and the goal to make television and radio services as universal and democratic as possible. This system depends on the government for protection against inequitable market forces (Hitchens 2006), but nevertheless can develop with a degree of autonomy from the government. This position was strengthened by other factors like the need to coordinate spectrum scarcity (i.e. fairly distributing the scarce airwaves for broadcasting), the maximization of political power (i.e. political parties that try to have the greatest influence on mass media), the drive to emancipate citizens (i.e. using the broadcasting media to foster citizenship) and the demands that the related area of telecommunication should be a universal service (Donders and Evens 2011). As a consequence broadcasting monopolies were set up in the UK and other Western European countries, mainly funded by public money. Although there is also a public service broadcasting station in the US, broadcasting there is typically organized as a commercial activity with very little government intervention. This was motivated by the 'free market of ideas' model in which the media sector would itself achieve a free flow of content and ideas by avoiding government intervention (Morris and Peterson 2000).

Although changes within the broadcasting systems are locally specific and have taken different forms in various parts of the world, depending on policy regimes, regulatory history, economic traditions, business conditions, viewer practices and cultural preferences (Tay and Turner 2010: 41), similar policy developments have taken place all over the world. In that respect Ellis's (2000) categorization is helpful, as it grasps a pattern of historical development that all broadcasting systems worldwide seem to have passed through (see also Chapter 5), moving from (1) an age of scarcity when broadcasting was – and was treated as – a scarce and therefore precious commodity, both in qualitative and quantitative terms, through (2) the age of availability when broadcasting proliferated and became largely available via increasing numbers of channels and programming hours (24/7), to (3) today's age of plenty when broadcasting is abundant, challenging the traditional frontiers of broadcasting policies.

BROADCASTING POLICIES IN THE AGE OF SCARCITY

Early broadcasting systems were characterized by monopolies and stretched from the post-war beginnings of radio and television (respectively World War I and World War II) until the end of the 1970s and the beginning of the 1980s. In a US context, this period has been labelled as the 'network era' (Lotz 2007) or the age of 'dial television' (Uricchio 2009). European communications scholars have denoted this era as the age of 'media within frontiers – managing scarcity' (Pauwels 2011) or 'defensive Europeanization' (Michalis 2007). On the whole, there was a substantial difference in the ways broadcasting was organized in: (1) Western Europe, where national public sector monopolies prevailed; (2) the US, where a few commercial companies took part in the broadcasting system; (3) countries with non-liberal regimes, where the state was in control of broadcasting; and (4) some countries in-between, such as Japan.

Apart from the UK, where in 1955 the commercial ITV network was launched, there was a limited commercial broadcasting market in the European area before the 1980s. In most European countries it was believed that the media and telecommunications sectors were so-called 'natural' monopolies that needed to be organized by the state, and assumed that the market could not operate in these fields. This policy option was further justified by analogue spectrum scarcity and very high infrastructural costs, and also deeply rooted in the belief that the media exerted a strong effect on the so-called 'masses' (Donders 2010). Hence, public service broadcasting was a means to establish a certain degree of government control over the mass media (first radio and then television) and thus over the masses. Besides controlling content, this could also result in the politicization of broadcasting organizational structures. Some or all levels of the broadcasting institution were, for example, subject to party political appointments, resulting in increased bureaucracy and a weak basis for competition (Burgelman 1989). But such competition and the market potential of broadcasting was not a priority for these governments, as there was no market, only a public. All this also meant that funding in the broadcasting industry was largely based on public subsidies through taxation, in the form of license fees or direct government grants. There was very little additional revenue from advertising and sponsoring. The audience was left with little influence and choice in programming. In addition, programming was oriented to satisfying the collective, public needs of the audience, rather than their individual, personal wants (Murdock 1993: 527; Shaw 1999: 153). Built on the strong conviction that mass media was powerful, modern liberal governments in post-war Europe defended the idea that public broadcasting could and should contribute to the development and preservation of liberal democracy, certainly after World War II. For this broadcasting had to inform, educate, elevate and empower 'the people'. This led to a distinct emphasis on education, information and 'high culture', omitting popular entertainment (see, among others, Van den Bulck 2001).

Radio broadcasting in the US had been organized on a commercial basis since 1919 and mainly happened through the airwaves. By the 1930s the three big commercial broadcasting networks, namely ABC, NBC and CBS, which would go on to be the major television players for three decades, already dominated the market. The period between the early 1950s and the 1980s was formative for the US broadcasting system, as the industrial norms of the television medium were then put in place (Lotz 2009). Many of the norms and conventions that were set in that period have persisted until now, despite substantial changes in the past years. Competition was chiefly between the three networks, which dictated the terms of production to the content producers (studios). The networks were also the only channels for broadcasting high-budget original programmes. Given the commercial nature of broadcasting, the funding was secured through 30-second spot advertisements, mostly sold in packages before the season began and based on crude audience information. The audience had no control over what, when or how they wished to see or hear the programmes, having to choose among undifferentiated programming options (Lotz 2007).

BROADCASTING POLICIES IN THE AGE OF AVAILABILITY

Between the end of the 1970s and the mid-1980s, the age of 'multi-channel transition' (Lotz, 2007) or the age of the 'remote control' (Uricchio 2009) dawned. Other scholars also refer to this period as 'media without frontiers – managing choice' (Pauwels 2011) or as the years of 'liberalization and re-regulation' (Michalis 2007). Until the mid-1990s or – depending on the world region – even the beginning of the 2000s, broadcasting was characterized by regulatory liberalization, with an exponential growth of television channels (see Chapter 5). The broadcasting industry also gradually incorporated emergent changes into its remaining standard operating procedures.

In Europe, broadcasting was liberalized and privatized during the 1980s, which led to the breakdown of the existing (state) monopolies. This liberalization happened on a national as well as on a European policy level, where advancing European integration had an increasing impact on broadcasting regulation in the various EU member states (e.g. the Television Without Frontiers Directive in 1989). Liberalization enabled the introduction of European private television and other commercial entrants (e.g. MTV Europe) into the EU broadcasting markets, accompanied by the rise of the cable and satellite industry. The latter development already foreshadowed the effects of the coming digitization of the broadcasting sector, with the end of spectrum scarcity enabling more channels to be licensed. Scarcity started to shift to abundance whereby private television was seen as necessary for augmenting consumer choice and autonomy. At the same time this led to the growth of non-European, predominantly US popular content (e.g. sitcoms, movies) on European channels. Declining technological constraints also opened the way for new broadcasting business models, such as subscription-based business models and pay-TV, which was introduced in those years (Donders and Evens 2011; Pauwels 1995).

Besides a surge in television channels and the entry of private television companies, the liberalization efforts in Europe led to an inflation of production costs, the growth of acquisition prices (i.e. significant price increases for buying television content like US series), the increasing importance of premium content (i.e. content that attracts and locks in a large audience like blockbuster movies or quality series) and the emergence of a 'dual broadcasting system'. The duality was shown in the differentiation between, on the one hand, the old public service broadcasters that relied chiefly on public funding and, on the other hand, the newly introduced and quite successful private broadcasters that obtained substantial revenues from advertising and sponsoring. Only pay-TV had an alternative revenue source of subscription income, but this did not become an immediate success in all European countries (Donders and Evens 2011).

In contrast to EU deregulation, the US government adopted additional regulation in the run up to and the beginning of the multi-channel transition. The networks were forced to relinquish some of their control over the terms of programme creation, in favour of the producers. However, 1980s US policy, epitomized by the 1984 Cable Act, was also characterized by considerable deregulation, which allowed massive consolidation and conglomeration in the broadcasting industry. Conglomerates gathered many television stations, broadcast networks, cable channels, production facilities and distribution facilities (like cable and satellite providers) into common ownership. This also led to a broadcasting landscape dominated by Multiple-System Operators or cable players, who could profit from the cost reductions enabled by economies of scale while simultaneously increasing subscriber fees for viewers. The misuse of this dominant market position was partly corrected by re-regulation through the 1992 Cable Act, but the cable and broadcasting technologies were changing extremely fast, especially as a consequence of digitization, convergence and the internet in the early 1990s. The various industry-specific regulations were no longer sufficient – new, all-encompassing legislation became necessary. This led to the US Telecommunications Act of 1996, which mandated digital transition, combined with deregulation. Regulators did not, however, use this opportunity to intervene in broadcast norms, by, for example, revisiting the vaguely defined mandate that stations operate in the 'public interest, convenience, and necessity' in exchange for using the public airwaves (Lotz 2007: 47; Lotz 2009; Mullan 2008).

As for viewers, during this period their influence over broadcasting content and control over when and how to consume television programmes increased significantly. First the number of channels multiplied, due to emerging (analogue) cable channels, new broadcast networks (e.g. Fox) and (advertising-free) subscription channels, which eroded the dominant position of the three main networks. Besides these changes in content choice, the viewers also gained more control by way of new technologies like the remote control device and the videocassette recorder (see also Chapter 5). Lastly, the methods for measuring audiences on the supply side grew in sophistication.

BROADCASTING POLICIES IN THE AGE OF PLENTY

We are now in the 'post-broadcast era' (Tay and Turner 2008: 71) or 'post-network era' (Lotz 2007: 7). Other names being used are the age of 'media beyond frontiers – managing abundance' (Pauwels 2011) and 'from TiVo to YouTube' (Uricchio 2009). This has been characterized (particularly in Europe) as a 'competitiveness, knowledge economy, and technological convergence' (Michalis 2007). The era in which we are now living is typically set from the mid-1990s or – depending on the region – even the beginning of the 2000s. We have arrived at a period where technological digitization and convergence are starting to reach maturity. This is, for example, shown in the spread of fixed and also mobile broadband, the multiplying of interactive digital television set-top boxes, the increase in mobile data use and the substantial diffusion of wireless technologies. Hence digitization is taking place on all levels, hardware, software and networks, while the converging consumer devices become increasingly powerful audiovisual media tools (e.g. smartphones, tablets). This also means that the ramifications of disruptive technological changes are becoming more visible, with a potential for mutually (re)shaping broadcasting policy and industry. The latter is coupled with changing habits and practices among user-audiences, which are showing themselves to be increasingly interactive and empowered (Mante-Meijer et al. 2011).

Although regulation is still often oriented towards separate vertical sectors, as shown by the different EU directives for television, cable and telecommunications, policymakers and regulators inside Europe are coming to grips with this new media environment and gradually acknowledging the technological developments within broadcasting media. A typical example is the revision of the Television Without Frontiers Directive of 1989, through the adoption of the Audiovisual Media Services Directive in 2007. The latter is an attempt to redirect policy from a vertical approach (as employed for traditional broadcasting) towards a horizontal, technologically independent approach to content regulation. The horizontal policy model implies that legal rules are no longer separated for the different sectors (with different rules for broadcasting versus telecommunications), but on the basis of the distinction between content production and transmission (Valcke and Stevens 2007). The Audiovisual Media Services Directive regulates the whole layer of content services, while other directives (on electronic communications) regulate the transmission layer. This means, on the one hand, that rules are technologically or platform independent: the rules are applied to all audiovisual media services, irrespective of whether these TV or TV-like services are delivered via the airwaves, cable, satellite, mobile networks, xDSL, the internet, etc. On the other hand, a difference is made between linear services, like television broadcasting, and non-linear services, like on-demand media. The first are services that are built on a linear schedule of programmes, the order of which the viewer cannot change, typically referring to conventional television services (via any form of transmission) but also webcasting and live streaming. The regulation for these linear services, still referred to as 'television broadcasting', is based on the former Television Without Frontiers Directive tier of rules, which have then been modernized and relaxed. Non-linear services, on the other hand, cover on-demand services where users/viewers are able to select the content they wish at any time, which includes examples such as video-on-demand or internet-based news services. These non-linear services are only subject to a 'light' tier of rules (Donders 2010).

These regulatory adjustments in the new media environment have been criticized for not being sufficiently sustainable, far-reaching, and consistent for all stakeholders (see Donders 2010; Valcke and Stevens 2007). For example, it is argued that the distinction between linear versus non-linear services could be difficult to uphold in the future, as the dividing lines between television, radio, the internet and mobile platforms (e.g. smartphones) are increasingly blurring. There is also no full technological independence as the rules only cover electronic audiovisual services, not the complete field for content within all existing online and offline media. The printing and publishing press are, for example, not subject to the Audiovisual Media Services Directive. Hence online news information from newspapers is more lightly regulated than the online news of television broadcasters, despite the fact that both are reaching and competing for similar audiences with the same advertising revenues. The changes in legislation also largely disregard two central stakeholders in the digitized and converged broadcasting landscape: users as content creators (e.g. citizen journalists) and the content distributors delivering audiovisual media services (usually edited by third parties) to the end-users. The latter typically refer to telecommunication operators using a digital television platform, offering packages of channels and services edited by broadcasters, production houses or other media companies. But it also includes new actors like social network sites (e.g. Facebook), online video services (e.g. YouTube), technology companies (e.g. Apple) and even search engine providers (e.g. Google) fulfilling the role of an intermediary or portal, providing a forum for users to make personal audiovisual content publicly available and filtering people with a specific profile to direct them to content of possible interest (Donders 2010; Valcke and Stevens 2007). Despite the problematic dimensions of policy noted above, European regulation has become increasingly important for broadcasting players in the post-broadcast era. European policy areas have also matured, in that more often than not they have become the starting point for changes at the national and local level (Michalis 2007).

We find a very different situation in the US, where government audiovisual policies – if any – have lagged behind the rapidly evolving technological and industrial changes, especially in the post-network era (Lotz 2009). Technological transformations, such as the digital transmission of television signals and the adoption of digital production technologies and audience devices, have reshaped the production, transmission and consumption of television and radio broadcasting significantly. In this digitized and converged media landscape, new portable – often extra-domestic – uses of television are unlocked (e.g. watching YouTube music clips on a smartphone; catching up with a series on a tablet; looking at live sports on a mobile television device), while there has also been an improvement in the visual and auditory quality of television (e.g. high-definition standards).

The steady stream of technological innovation has undermined long-established industrial practices, leading to a new set of conventions. For example, today multiple strategies coexist to enable content providers to get advertising funding. This was spurred on by the rise of internet advertising possibilities and the growing concern about new recording systems like the digital video recorder (DVR), which made ad-skipping possible. The range of advertising strategies in the post-broadcast era includes 'old-fashioned' forms of single sponsorship and various types of product placement, as well as new forms of 'branded entertainment' that inextricably integrate the advertising message with the content (e.g. infomercials). The broadcasting industry has also been experimenting intensely, and successfully, with various forms of subscription and transaction financing (e.g. pay-per-view) that eliminate advertisers. Such forms of financing have had an influence on programme content itself, in terms of making it more risky, experimental and innovative and less mainstream and common than was the case in the era of spot-advertising and channel scarcity.

Another example of changing industrial norms is the fracturing or disaggregation of the previously dominant distribution practices. Even more striking than the large extension of channel options during the multi-channel era, we have entered a phase with a plethora of audiovisual content outlets. The business of content creation is being reconfigured, with the more narrowly targeted channels requiring cheaper production techniques, and with the release of previously produced programming through a large array of digital outlets. For example, certain programmes are syndicated to enable viewing via on-demand services on DVRs or catching up via an app on a tablet computer. The former technological bottlenecks in broadcasting are gradually disappearing, shifting the logic from one of scarcity to one of abundance (Anderson 2009).

As noted, on the US policy front the logic of commerce is foremost in shaping the regulatory mechanisms. There is only minimal governmental intervention in US broadcasting regulation. Given the deregulatory policies followed since the beginning of the 1980s, the already modest touch of government involvement has become even lighter. This has led to the consolidation of stations into fewer – less local – hands and to a significant conglomeration of media industry ownership. The only substantial government initiative in broadcasting is the mandated shift to digital signal transmission. However, this particular development illustrates the ineffective and haphazard character of government intervention in the operation of broadcasting. Although the Congress mandated the digital transition in the Telecommunications Act of 1996, the matter was only brought to completion – 13 years later – in June 2009. What is more, the functional reinvention of broadcasting, with the reallocation of spectrum, has not been used as an opportunity to restate or clarify the responsibilities of broadcasters in exchange for their free use of the public spectrum.

Reflection: Rethinking US Policy in the Post-Broadcast Era

Think about how policy in the US could have possibly used the functional reinvention of broadcasting – the moment when spectrum was reallocated – for the good of society and the economy. How could policy have been different? Which regulatory measures could have been beneficial?

Chapter Summary

■ This chapter presented an overview of the historical development of broadcasting as a social institution and a 'cultural form'. Being omnipresent in the lives of many people all over the world, we investigated how established and accepted meanings, systems and regulations of broadcasting have been challenged and renewed throughout history. We started with the observation that radio and television have always had a strong connection with the domestic sphere (domestication), in terms of ubiquity, familiarity, everydayness and ordinariness. However, this was not inscribed as such in the 'nature' of broadcasting. The dialectical relationship between the private and public domains of everyday life, and the reconfiguration of both domains were essential to understand the renewability and resilience of broadcasting over time.

■ We then looked at broadcasting and the changing conceptions of the home. The media home has historically been built upon three cultural metaphors through which domesticity and media have been imagined: theatricality, mobility and sentience. These are apparent in three different media housing types: the 'home theatre', the 'mobile home' and the 'smart home' of the digital future. We observed how each of these metaphors is coming back in the contemporary age of digital broadcasting.

■ The chapter next dealt with consecutive models of broadcasting systems. Like media technologies, broadcasting systems are also the historical result of different industrial, economic, cultural and regulatory processes. The two main opposing models are systems based on government funding and intervention (for example, in the UK) and systems conceived of as a commercial activity relying mainly on advertising revenues (for example, in the US).

■ The historical development of broadcasting systems has gone from (1) the age of scarcity when broadcasting was a scarce and thus precious commodity (from the introduction of radio and television until the late 1970s), through (2) the age of availability when broadcasting proliferated and became more widely available (from the late 1970s until the mid-1990s), to (3) the age of plenty when broadcasting is abundant, while challenging the traditional frontiers of broadcasting policies (from the mid-1990s until now).

3 THE BROADCASTING INDUSTRY

Traditionally, the broadcasting industry is subdivided into a 'hardware' sector, comprising cables and middleware technologies such as cable modems and television sets, and a 'software' sector, which refers to the so-called media content industry (MCI) (e.g. television programmes, radio formats, advertising content, etc.) (Pauwels 1995). Both these sectors can be looked at from a production and distribution perspective. Production refers to all technologies that are necessary for the producing (e.g. movie cameras, recording material), transmitting (e.g. satellites, coax cables) and receiving (e.g. radio devices, antennas, set-top boxes) of audiovisual content. Distribution covers the different ways of organizing and standardizing the transmission and reception systems for software (e.g. technical requirements for Digital Video Broadcasting, High Definition TV standards), as well as the means for enabling the distribution of hardware (e.g. training, know-how for efficient implementation).

Each of these different activities traditionally implied stable, clear-cut roles and was represented by distinct stakeholders. However, the techno-economic, political and regulatory changes that are instigating digitization and convergence are shaking up the traditional roles and stakeholders within the broadcasting sectors, and traditional media-specificity, which used to be the organizing principle in the media industry, has been abandoned. As a consequence the market and business structures in the broadcasting industry are being reshaped. Basically, the so-called silo-structure, which refers to the media- and sector-specific organization of the broadcasting industry, is being eroded (Donders 2010: 423; Donders and Evens 2011: 24). Indeed, in the post-broadcast era audiovisual content is no longer exclusively linked to a specific platform and to a particular revenue stream (see also Chapter 2).

Clearly, the broadcasting industry and related regulation are certainly not fully transformed in the West, and even less in other parts of the world (Tay and Turner 2008, 2010; Turner and Tay 2009). However, digitization and convergence are bringing about some critical changes in the business environment for broadcasting. It is argued (and sometimes already demonstrated) that this will lead to a multi-platform expansion of the media sector, requiring adaptation of media management towards a so-called 360-degree approach. This novel approach hatches new ideas for content in the context of a much wider range of distribution possibilities than linear television alone (e.g. mobile delivery technologies, interactive games, etc.) (Doyle 2010) (see also Chapters 4 and 5). In order to understand how the broadcasting industry is evolving in the post-broadcast era, this chapter will deal with how the industry is organized and how business economics are changing in relation to key technological changes in distribution, transmission and reception.

Reflection: Digital Erosion of Borders

Think about examples where you see content being freed from specific platforms or devices, and thus from the related revenue source. In what way does this create opportunities and risks for industry, policy and users?

CHANGES IN COMPETITION

One of the most crucial transformations that broadcasting markets and firms are going through as the result of digitization is the adjustment of their strategies to maintain, regain or pursue a competitive edge, which implies finding the optimum match between an organization and the competitive environment (Küng et al. 2008). Building upon Porter's (1980) classic analysis, we can speak about five basic factors or forces that configure competition in a sector and, in that way, co-determine the performance of companies, that is: (1) potential entrants; (2) possible substitutes; (3) the power of buyers; (4) the power of suppliers; and (5) the intensity of rivalry among existing firms. Porter argues that the competition will be fiercer when the bargaining power of buyers or suppliers is larger (factors 3 and 4), or when existing products or services can be easily replaced by similar substitute goods (factor 2). A market where newcomers can easily enter because of low entry barriers will also be much more competitive (factor 1). The combination of all five factors determines the ultimate profit potential in the industry, which is measured in terms of long-term return on invested capital.

Within media industries and broadcasting organizations in particular (see, among others, Van Thillo 2011), it is assumed that digitization is strongly altering these forces. First, the legal and financial barriers to enter the market are increasingly being lowered. This process, which first took place as a result of liberalization and privatization policies (see Chapter 2), is today accelerated through digitization. The increasing capacity for more distribution channels has definitely lowered the legal and financial barriers for the entry of new players and resulted in more competition. However, it is still very costly to produce successful commercial television with a mass reach (see also Chapter 5).

Second, the substitutes for traditional television broadcasting have also changed over the years. While substitutes for end-consumers were at first very limited, with only VCR and pay-TV in the 1990s, the possibilities for receiving news and information, and enjoying entertainment and culture, have increased enormously in other ways (Lotz 2009). The internet, social media, games, personal (or digital) video recorders and interactive television services have become important delivery platforms for audiovisual content. Meanwhile, the substitutes for advertisers have evolved drastically. First, advertisers exchanged television for the less-strong advertising channels of magazines and video. By the end of the 1990s there was the internet-hype, where online advertising was expected to largely take over the role of traditional broadcasting (Doyle 2002: 54–7; Jaffe 2005). However, ten years later, the role of the internet has been put more into perspective, and broadcasting (especially television) has regained its strong position in the display advertising market (UBA 2014: 26).

Third, the process of digitization is also affecting the bargaining power of suppliers, buyers and other stakeholders, such as advertisers (Aris and Bughin 2005; Cauberghe and De Pelsmacker 2006; Napoli 2010a). In analogue times the content providers had little power, as there were often only a few networks and broadcasting channels to which they could sell their programmes. In many countries, there was only one public service broadcaster on the air. This changed towards the end of the twentieth century, when these content providers could set up competitive auctions for output deals. A decade later, the system is again more balanced, with more content available, an audience market becoming more settled, and broadcasting companies regaining some of their bargaining position in the advertising market. Hence, the power of the advertisers, which initially increased with the belief in a surge of online advertising, has decreased as television still appears to be a powerful advertising medium in many cases (Pfeiffer and Zinnbauer 2010). The distributors, mainly cable companies, telecommunication operators and satellite companies, have had to give up some power. At first they had a distribution monopoly, but this changed when new distribution technologies became available for consumers (Van den Dam and Nelson 2008). Especially in recent years many consumers have gained access to additional channels for receiving radio and television, in particular digital platforms (e.g. IPTV, Hulu on the internet). Nevertheless, we see that in several countries the distributors still hold a very powerful position, depending on the regulatory context. The end-consumers have had little to no bargaining power in the broadcasting business. In general we see that the revenue streams in the broadcasting landscape are becoming less strictly divided between advertising versus public funding. New income streams are becoming more available, such as direct payment by consumers (e.g. subscription) and business-to-business exchange fees (e.g. transmission fee) (Moran 2005).

Finally, the competitive rivalry within the broadcasting sectors, both hardware and software, has also changed in the past 20 years. These changes can be attributed to market players having gained in strength, such as new network stations (e.g. Fox in the US) becoming popular or public service broadcasters (e.g. the BBC in the UK, VRT in Flanders-Belgium) becoming a stronger challenger.

FROM VALUE CHAINS TO VALUE NETWORKS

The concept of a 'value chain', an analytical construct introduced by Porter (1985), and originating from the fields of industrial organizations and microeconomics, is helpful in identifying how the strategies and the competitive environment within the broadcasting industry have been influenced by digitization. The basic principle is that organizations employ a variety of resources to create products and services for the market. The surplus value that exceeds the costs of the resources represents profit. The activities of companies can be disaggregated into a series of processes, going from the supply to the demand side, in which particular sets of roles and tasks are necessary to create and distribute products and services. Each of these processes is assessed from the perspective of the 'value' it adds.

Before the digital era there were relatively straightforward value chains in the (mass) media and broadcasting industries. Traditionally, the consecutive stages were: (1) developing content; (2) packaging content; (3) distributing content; and (4) reception of content. Each stage involved distinct actors like television production companies (e.g. Endemol) (1), television network companies (e.g. Fox) (2), cable operators (e.g. Comcast) (3), and the audience (e.g. television viewers) (4). With digitization it has become difficult to think about value chains in linear terms, as value chains have become much more complex and expansive, and traditional boundaries among the actors involved in the four stages have blurred as a result of mergers and acquisitions between old and new media. In order to better encompass the increased complexity of the broadcasting industry in the digital era, we need to extend the notion of a value chain to the less linear notion of a 'value network'.

A value network describes sequences of value-adding activities that make clear how companies within the boundaries of their industrial activities mobilize their resources in relation to surrounding actors. Hence a value network is a set of relevant activities behind a product or service offering. The principal nodes in the network are not the actors performing the activities, but the activities themselves. The relationship between these nodes can consist of flows of information, services, materials or financial resources (Ballon 2005).

A value network consists of three major constituents. First, business actors are physical persons or companies that participate in the creation of economic value through the mobilization of tangible or intangible resources (i.e. financial, intellectual and social capital) within a business value network. Second, business roles are logical groups of business activities that are fulfilled by one or more actors. Business actors provide value to and/or derive value from the business roles they play. For example, a customer, supplier and transportation company all play business roles in order to fulfil business processes. Third, business relationships can be the contractually defined exchange of products or services, financial payments or other resources (Ballon 2005; Donders and Evens 2011).

From an economic point of view, the digitization of the broadcasting industry asks for a value network approach, as it is no longer a case of positioning a fixed set of activities along a value chain (Normann and Ramirez 1993). Rather the focus of strategy is exceeding the company or even the industry itself and is oriented to the whole of the value-creating system. In this system, the different economic actors (like suppliers, partners and customers) co-produce value. The main strategic challenge is to reconfigure roles and relationships among this constellation of actors to improve the fit between a firm's competencies and the customers' needs.

BUSINESS MODELS

The changes in value networks can further be linked with business model analyses. In a business model, technical, organizational and financial aspects are combined in order to capture the complexity of new business structures. Business models can be defined as: 'a description of how a company or a set of companies intends to create and capture value with a product or service. A business model defines the architecture of the product or service, the roles and relations of the company, its customers, partners and suppliers, and the physical, virtual and financial flows between them.' (Ballon 2005: 8)

By defining business models in this way we do not look at one single firm but at the entire network of actors involved in the production, distribution and consumption of a product or service. In that way it reflects the increasing complexity of innovation processes and enables us to include the value networks that have become more complex and more expansive.

Reflection: Identifying Business Models in Broadcasting

Think about what kind of business models you know or observe in the broadcasting industry, as for example for television stations that do not get public funding but depend on advertising. What kind of actors are involved? Which roles do they play? How are the relationships between the roles enacted?

DIGITAL BROADCASTING INDUSTRY FROM A VALUE NETWORK PERSPECTIVE

The traditional value chain of broadcast media in the network era was organized very simply and straightforwardly, also indicated as the vertical supply chain (Doyle 2002). This comprised three main stages: (1) acquiring or producing content (programmes); (2) packaging content (i.e. scheduling programmes into channels); and (3) distributing the packaged content to the audience (transmission) (Küng et al. 1999: 35). A typical broadcasting organization (e.g. television channel) controlled all of these stages, including the national broadcast network that they either owned or to which they had guaranteed access.

Since then the simple linear value chain has significantly expanded, which is reflected in the increasing number of production means, channels, transmission modes, consumer devices, audiences, funding options, etc. Hence, the different chain stages, representing bundles of fully integrated processes, are being 'unbundled' into separate stages or business roles. As a matter of fact, the new value network can also comprise activities outside the traditional media and broadcasting sector. A typical example is users that generate, upload and/or rate content themselves and in that way add value to online video platforms. Hence, as Figure 1 below demonstrates, the generic value network of broadcasting in the post-broadcast era looks much more complex, given the substantial expansion of business roles inside and outside the sector (Donders and Evens 2011). The value network is vertically organized into different industrial domains somehow linked to the broadcasting industry (in the columns separated by dashed lines): the content industry, the content distribution industry, the internet industry and the software industry. For the sake of clarity we have left out the telecommunications industry of fixed and mobile telephony. However, the latter have been instrumental in the breakthrough of digital television uptake in many countries, in the way that these fixed and mobile telephone services were bundled with digital television and internet services, in 'triple play' (television, internet and fixed telephony) and 'quadruple play' (television, internet, fixed telephony and mobile telephony) offerings.

The generic value network is also organized horizontally according to different services delivery phases (in the horizontal dashed rectangles). These refer to the way in which the business roles deliver services to each other, often (but not always) chronologically from top to bottom. The main business roles in the digital broadcasting industry are:

(1)   Content creation and production;
(2)   Content aggregation and packaging;
(3)   Content distribution;

(4) Content service provision;
(5) Content service and technology consumption.

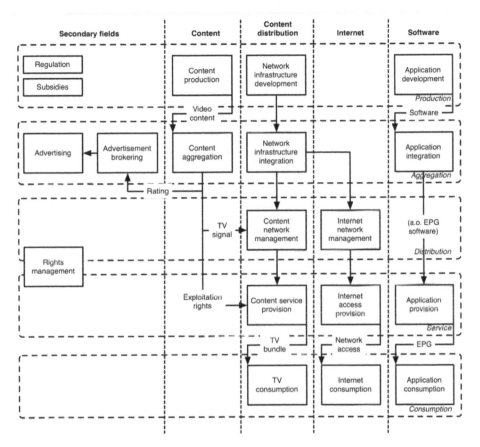

Figure 1: Generic value network for (digital) television and distribution (Donders and Evens 2011: 30)

Clearly, all stakeholders have their own stakes in the digitization process. It can be argued that in order to fully understand the fluidity and complexity of the changes the broadcasting industry is going through, a horizontal perspective is more helpful.

CONTENT CREATION AND PRODUCTION: CONTENT AND APPLICATION DEVELOPERS

A central activity in the broadcasting industry remains creating and producing content, as performed by companies like Warner Bros., Universal, Disney or Endemol. Studies show that this is not changing in a digitized and converged media landscape. On the contrary, the content segment has enormously increased its economic leverage, which reflects its importance for attracting consumers, as Bill Gates' dictum 'content is king' suggests (Küng et al. 1999; Van den Broeck 2010). In addition, in an 'unbundled' value network, the content developers have more options than ever to monetize their creative products. In the age of scarcity (see Chapter 2), content producers either largely resided within the broadcasting companies, or the channels and networks were their only (and therefore powerful) customers. During the age of availability (see Chapter 2), the number of possible buyers increased. Yet in the current value network of the post-broadcast era they can choose between different pathways to bring their products to target audiences. They can still sell them to one of the many content aggregators (i.e. channels). However, if they are a powerful player in the market, they also have the option to jump over the aggregator, and go directly to the content service provision managed by the distributor. Or they can even think of setting up their own dedicated channel, to be broadcast via one of the managed content distribution systems or directly via the internet.

Case Study 3.1: ESPN

The sports entertainment channel and content producer ESPN (Entertainment and Sports Programming Network), established in 1978, is 80 per cent part of the Walt Disney media conglomerate. After various owners it was majority-bought by ABC/Capital Cities in 1984, which Disney acquired in 1995 for $19.5 billion. In fact, Disney wanted the US broadcast network ABC, not ESPN. However ESPN is now said to be responsible for 40 per cent of Disney's operating income, 60 per cent of its free cash-flow and as much as half of its share price (*The Economist* 2013). In 2009 Disney's revenue was $36.1 billion, the highest of all global media companies (Winseck 2011). ESPN was one of the first channels to offer live video on its website and it has launched an application so that viewers can stream the channel on mobile devices (www.espnplayer.com).

ESPN is a good illustration of how a traditional (analogue) broadcast channel can transform itself and become a profitable and powerful player in the digital value network, although it needs to be said that the conditions were favorable, since fans prefer to watch sports live, which means that fewer advertisements are skipped by way of fast forwarding. Also, the rights to broadcast certain games are often exclusive, which means viewers cannot see them elsewhere. Finally, ESPN was the pioneer of 'affiliate fees', i.e. the fee cable operators have to pay for the right to carry each network. In 2013 it was estimated that ESPN would earn $6.6 billion from affiliate fees, which would be more than three times what it generally made from advertising. Thanks to the many exclusive sports rights, the sports channel has been able to raise its fee to $5 per subscriber, per month. This is far higher than any other network's fee (*The Economist* 2013).

The creation of audiovisual content often requires the interaction between creative crew, technical crew, production companies, broadcasters and other possible parties (Guiette et al. 2011). Sometimes the idea is initiated by the broadcaster or content aggregator in order to develop a specific programme. The latter then commissions a specific production company. In another creation process the broadcaster aims to develop a certain concept and then requests one or more production companies to prepare an offer, from which the broadcaster chooses. The creation can of course also be initiated by a production company, which then tries to sell the concept to a broadcaster, with which they want to develop it further. The old-fashioned process of in-house production by the broadcaster also still occurs, in particular in the case of public service broadcasters for those programmes that belong to their core assignment (e.g. news, culture, education). Finally the simplest and – most often – cheapest way is to buy programmes on the international trade market (e.g. programme franchises, see also Chapter 4).

The audiovisual content itself can be very diverse along a wide range of different genres and forms. Hence, the production sector is fragmented and often specialises in particular types of content. In an increasingly global and commercial broadcasting landscape this kind of specialization is often translated into intellectual property rights. Generally, there is a heightened awareness and concern about how to safeguard and control content-related ideas, which has resulted in the formalization of ownership under the protection of property laws. In particular in the international television market this is shown in the development and trade of 'formats', i.e. a set of invariable elements in a programme out of which the variable elements of an individual episode are produced (Moran 2005: 292–6). Specific applications of formats have also for some time been discussed in the context of radio (cf. Johnson and Jones 1978). Typical examples of such formats are *Big Brother* and *The Voice*, both developed by the Dutch production company Endemol, which is one of the companies at the forefront of developing and selling successful formats.

Reflection: Formats in Broadcasting

Think about your own national, regional or local television programmes. Which programmes are based on existing formats? Are there also international formats that have been bought? Which of those formats are from your own country or region? To what extent have they been exported to other countries?

Obviously, the adaptation of programme ideas to another national or cultural context is not novel. In the past, programme templates from the US and UK were replicated in many parts of the world. Adapting a popular programme enabled a local producer or broadcaster to lower the risks, as the template had already withstood two tests: trialling before broadcasting executives and airing before viewing audiences. However, in recent years the once casual and spontaneous process of international adaptation of broadcasting programme ideas has been consciously routinized and formalized by way of formats. The latter include value-adding elements (like the 'format bible'), format-marketing arrangements (industry festivals and markets), licensing processes and self-regulation administered by the industry association FRAPA (Format Registration and Protection Association).

Interestingly, in the post-broadcast era the notion of content is no longer a synonym of television programmes and programming alone, but it now also includes the creation of new sequences of image and sound, availing and engaging in interactive services and accessing data and information. The television set now incorporates many functions besides television broadcast programme reception and channel zapping. Playing DVDs, streaming programmes and off-air recording allow time-shifted viewing. The new television landscape has also brought to the television screen Electronic Programme Guides (EPG), internet-related services and email, games, information services (e.g. e-government), teleshopping and many other applications (Moran 2005: 293). The growing influence of ICTs on the media sector entails that the terms 'software' and applications are increasingly being used interchangeably with that of 'content' (Küng et al. 1999). Consequentially, this means that the role of software and application development also becomes important in the value network of the digital broadcasting industry. In addition, this increases the possibilities for (digital) tracking and profiling of broadcasting audiences in their engagements with their enhanced and (internet) connected television set. Yet these increased means of surveillance and data-mining have also led to new concerns regarding privacy and data protection (Burke 1999).

Case Study 3.2: Electronic Programme Guide (EPG)

While in the analogue broadcasting age of scarcity finding a programme was easy with the help of a newspaper and only a few broadcast channels to choose from, this has changed in the digital broadcasting age of plenty (*The Economist* 2010). Now the most common way to select the channel in the digital offer is via the Electronic Programme Guide (EPG) on the Digital Video Recorder (DVR). This means that the application software for the EPG interface can have a powerful influence on how the selection from several hundred channels – often combined with an offering of pay-per-view films and video-on-demand options – happens. The software increasingly entails recommendation algorithms that can suggest possible preferences for the viewer, based on former viewing behaviour.

This has led to new roles and actors in the broadcasting value network. Sometimes the EPG application is bundled together with the hardware device (set-top box or smart TV), as for example with TiVo, Apple TV or Google TV. In other cases there are separate companies that specifically develop this type of software, like Rovi Corporation, Cisco Videoscape or Zappware. In addition, television with a built-in internet connection (i.e. connected television sets by e.g. Samsung, Sony or Vizio) also offer their own EPG interfaces to surf the extended number of digital channels and television options.

CONTENT AGGREGATION AND PACKAGING

Most often content aggregation involves the bundling of internally produced and externally acquired content products under a media brand name (Donders and Evens 2011). In broadcasting, this kind of aggregation is typically done by actors like broadcast stations (e.g. ABC in US, BBC in UK, ZDF in Germany, VRT and RTBF in Belgium), cable networks (e.g. NFL, MTV, ESPN), satellite channels (e.g. Al Jazeera, CNN), pay-cable or premium channels (e.g. HBO, Showtime, Cinemax, Prime), video-on-demand (VOD) channels, pay-per-view (PPV) channels, on-demand PPV, digital music channels, (print) media companies (venturing into broadcasting) (e.g. NGC National Geographic Channel, Sanoma media) or video portals (e.g. YouTube, Vimeo, Hulu).

In traditional broadcasting, the separate roles of packaging, scheduling and integrating content were closely linked and often even taken up by one actor, most often the (public or private) broadcaster (see also Chapter 5). In the digital era the trend points towards 'unbundling' packaging into a distinct stand-alone stage. A typical early example of this was MTV. The business model of this music channel was based on acquiring promotional videos from music publishers, packaging these into a channel, and then selling this one media brand on to distributors. The new media competence of media aggregation involves so-called 'enhancing', which is reworking acquired content to fit specific niche audience requirements. In this way packagers are essentially intermediaries, in the core based on building alliances in three directions: with content providers, content distributors and consumers (Küng et al. 1999).

CONTENT DISTRIBUTION

In the traditional analogue television system, all stages, from production via distribution to reception, occurred in an analogue way. This meant that programmes were recorded as analogue, processed as analogue, stored as analogue, transmitted as analogue and the audience received the signals in an analogue way. There were three main analogue network infrastructures available: the terrestrial network (through the ether spectrum), the coaxial cable (CATV) network and the satellite network. Specific to terrestrial television is the fact that the available (public) spectrum is limited, which meant that specific regulations for these signals had to be put in place in order to fairly divide the scarce frequencies. As a result, the number of broadcast channels that can be received via the analogue terrestrial network is limited, in contrast to cable and satellite reception where this is less so. On the receiving side of each of the networks, different equipment is required and it needs to be coupled with the television or radio. In order to receive the terrestrial airwaves an antenna is needed (e.g. indoor or on the house). For CATV the television set need to be connected with the cable (possibly via a cable modem), and for satellite a satellite dish is required. In general, the reception via cable and satellite networks were of a better quality than analogue terrestrial reception. The most commonly used systems have varied from country to country and have grown historically.

The analogue television signal consists of lines and frames. When black and white TV was standardized in 1948, the US adopted a system with 525 lines of pixels. Europe introduced a system where an image is formed with 625 lines (576 visible) and 25 frames per second for the visual perception of movement. The European and American systems were and still remain incompatible. During the introduction of colour television, a similar dispute raged over standards, again leading to incompatibility. In 1953, the Federal Communications Commission (FCC) established NTSC (National Television Systems Committee) as the standard for US colour television. The Japanese followed the American standard. Europe judged that the quality of NTSC was not optimal and started developing different norms throughout the continent. France developed SECAM (Séquentielle Couleur Avec Mémoire), which was later adopted in Eastern Europe and Greece, while Germany created PAL (Phase Alternated Line), establishing a norm for Western Europe, Brazil and Southern Asia (Alencar 2009). Historically, it has often been shown that the best standard from a technological point of view did not necessarily become the most widely accepted (Walravens and Pauwels 2011).

DIGITIZATION OF TELEVISION DISTRIBUTION

The infrastructural transition to digital television can – from a technical perspective – be seen as an inevitable evolution of analogue television. At first the production process was digitalized, but transmission and reception were still analogue. With the transition to digital television, the latter processes are also digitized. This means that the analogue signal is coded into digital bits and bytes. The result is – in principle – a better quality of sound (with stereo sound) and a higher-resolution image (Alencar 2009).

Enhanced quality of image is certainly the case for high-definition television (HDTV). Where standard-definition television (SDTV or non-HDTV) refers to the common resolution for broadcasting digital television, HDTV can also be digital and provides even better quality, as the resolution of the image is higher. The first concrete HDTV initiatives had already been taken in the 1960s by Japan, in particular by the Japanese public broadcaster NHK (Nippon Hoso Kyoka) (Pauwels 1995). Europe tried to counteract the Japanese initiative with its own industrial HDTV project Eureka 95 in 1986. However, the latter initiative largely failed, for a number of reasons. The project only involved the European consumer electronics industry (like Philips, Thomson, Bosch, Thorn and EMI), and lacked sufficient other stakeholders in the value network (like content providers, (public) broadcasters, non-EU companies, consumer organizations, telecom companies and others). In addition it was only meant for analogue transmission, while the US in the end decided to develop an all-digital HDTV system (Walravens and Pauwels 2011).

HDTV now comes in three main types of screen resolution (720p, 1080i and 1080p) that provide similar improved image quality in comparison to SDTV (480i). Progressive (p) or interlaced (i) refer to the scanning method utilized. In progressive scanning, all pixels are refreshed at the same time, while in interlaced scanning, the equal and unequal lines are refreshed alternately. This means that progressive is better for fast-moving images, but interlaced requires less bandwidth. Bandwidth is important as the bit rate of one HDTV programme allows the broadcasting of four SDTV programmes (Alencar 2009). Since the mid-1990s the Japanese public broadcaster NHK has been working on the next step in the standardization of TV screen resolutions, which would be Ultra High Definition Television (UHDTV) or Super Hi-Vision (SHV). The resolution would be 16 times sharper than an HDTV image, and would also be backwards compatible (Walravens and Pauwels 2011).

For digital television to be interactive, local storage of information and/or a return channel to provide interactivity are needed (depending on the service). Digital television sets already exist that provide interactivity via built-in transcoders. These devices are called 'integrated digital television', meaning that they are completely digitized. For regular television sets, a set-top box (STB) needs to be connected to the TV-set to access digital TV and interactive services. The STB is a decoder that receives the digital television content and converts it for an analogue television. By means of the return channel, the viewer can also access interactive services. Many interactive services come with a cost. Access or subscription to these services can be verified via a smart card that is inserted into the set-top box (Van den Broeck 2010).

The transmission of digital television is also standardized in six main global standards: (1) the European standard DVB (Digital Video Broadcast); (2) the US standard ATSC (Advanced Television System Committee); (3) the Japanese standard ISDB (Integrated Services Digital Broadcasting); (4) the Brazilian standard ISDTV or ISDB-Tb (International System for Digital TeleVision); (5) the Chinese standard for digital terrestrial broadcasting DTMB (Digital Terrestrial Multimedia Broadcasting); and (6) DSL (Digital Subscriber Line), which is the technological standard to upgrade the twisted pair copper line network, originally only used for Plain Old Telephone Services (POTS), for high-speed data transfer of voice, internet and television (i.e. Internet Protocol Television IPTV).

DIGITIZATION OF RADIO DISTRIBUTION

The digitization of radio broadcasting has been and remains a road with many hurdles. The technological push towards digital radio has not (yet) been successfully picked up by consumers, despite the large investment by policy and industry. There exists a kind of chicken–egg situation where producers are not willing to produce devices and programmes as long as listeners are not prepared to adopt them, while listeners are not inclined to adopt (expensive) devices while there is too little supply or choice. In addition, with the rise of the internet, the notion of 'radio' itself is being redefined, with new kinds of distribution and reception of music and other audio programmes.

The traditional FM (Frequency Modulation) radio was introduced in the US at the end of the 1930s, when AM (Amplitude Modulation) radio broadcasting had already become a mass medium. The first enhancements of FM radio came with the addition of stereo in the 1960s. The first steps towards digital radio broadcasting were taken in the early 1970s when European broadcast engineers began to study the possibilities for improving analogue broadcasting with digital technology. First RDS (Radio Data System) for textual information (e.g. traffic information) was developed and introduced in the late 1980s. Comparable to teletext for television, the basic idea is to use the extra capacity of the analogue system for providing additional digital services (mainly textual data) for an audience with suitable broadcast receivers. In the trade and broadcast industry press, as well as in media research, notions like 'convergence' and 'information society' became prominent key issues. However, this had little immediate impact on research and development relating to digital broadcasting systems in Europe and the US. The engineers were focused on network digitization, crafting separate media-specific digital systems for radio and for television, instead of integrated solutions (Ala-Fossi 2010).

In this way DAB (Digital Audio Broadcasting) came into existence, as the outcome of a large European R&D industry project (Eureka 147). This was (together with the Eureka 95 project on HDTV) initiated as a kind of European industrial policy counter-attack against the Japanese dominance in the consumer electronics industry, especially with the Japanese-American HDTV system being developed and proposed as the new world standard for television. The European project group (with Philips, Thomson and Bosch in the core as private companies) believed the new DAB system would be adopted worldwide and thus give the European electronics industry a competitive advantage (Ala-Fossi 2010; O'Neill and Shaw 2010). DAB was technologically radical in that it could not only deliver sound, but with good mobile reception it could deliver any sort of data (e.g. text, pictures, slideshows, video clips, web pages). In addition it was possible to deliver five CD-quality channels or nine near FM-quality channels on one frequency. Later improvements (DAB+) made it possible to deliver 28 near-FM-quality channels. Despite the fact that more than 285 million people around the world could receive around 550 DAB services in 2003 (Oxera 2003) and that DAB was implemented in 28 countries in 2005 (World DAB, 2005), the take-up and adoption rate of the new radio devices has stayed low in most countries. One of the main reasons is the limited added value from a user perspective. Studies show that the main reasons for not having a DAB set are "no need for the service" (59 per cent), followed by being "satisfied with existing services" (39 per cent) (Ofcom 2013a: 23). The largest DAB market was the UK with over seven million receivers (Plunkett 2009; World DMB 2009). The DAB digital radio set take-up is also highest in the UK, with 48 per cent of radio listeners in 2013, which is far ahead of all other countries (with around 20 per cent or less) (Ofcom 2013b: 192). The Bosch Group continued on its own with DAB multimedia and developed a DAB-based system called Digital Multimedia Broadcasting (DMB) that could also deliver video for mobile reception.

In order to digitize AM broadcasting, a group of broadcasters and manufacturers formed the Digital Radio Mondiale (DRM) consortium in 1998. This consortium developed a new DRM standard as a modular innovation that could run on the existing AM channel allocation and where broadcasters could keep their own trans-mitters and networks. In that sense, from a social and economic perspective, it was much less radical and demanding. A further upgrade to DRM+ even enabled digital FM broadcasting (Ala-Fossi 2010).

DAB has not become the worldwide system of digital radio, partly because the US commercial radio industry saw it as a threat to their vested interests and in 1991 started to develop an alternative digital radio system under the umbrella of a single development company, iBiquity Digital Corporation. The latter proprietary system was called IBOC (In-Band, On Channel) because it used the already allocated AM and FM radio wavebands and the present channel allocation. This standard, later marketed as 'HD Radio', originally enabled the simultaneous broadcasting of a standard analogue signal alongside one near CD-quality digital signal, and a small amount of additional data. As this system also requires people to buy new receivers, it remains to be seen if it will make a significant breakthrough in the US.

Other systems for digital radio broadcasting were also developed. In Japan the ISDB was designed to be a converged platform that could offer digital television (ISDB-T One-Seg) as well as digital radio (ISDB-TSB Sound Broadcasting), together with additional digital services, on fixed and mobile (handheld) devices. Despite the fact that it was an open standard, so far only Brazil has adopted ISDB-T for its digital television standard. In Europe, the DVB standard was extended with DVB-H, a relatively cheap and efficient high-data downstream channel to mobile handheld devices, enabling new kinds of video, audio and data subscription services. This system has had modest success in Europe, mainly in Italy and Finland. In South Korea, the European DMB standard was enhanced and updated for mobile TV and radio (Ala-Fossi 2010; Alencar 2009). There are also the digital satellite radio systems. Around 1990 Digital Satellite Radio (DSR) became the first technology of digital transmission via satellite, in particular in Germany. However DSR failed the market test and did not survive. Also the ASTRA satellite company, a major satellite broadcasting company in Europe, started to offer Astra Digital Radio, albeit with limited success. In the US, originally two companies, XM (with more than 170 stations) and Sirius (with more than 130 stations), offered packaged programming as pay-audio, to a large extent for truckers and cars moving in rarely settled regions not covered by terrestrial signals (Kleinsteuber 2011). These two satellite radio providers had more than 16 million subscribers at the end of 2007, which is almost twice as many as two years before (Thierer and Eskelsen 2008). In that same year both companies announced that they would merge to Sirius XM, which has since extended its services to online and mobile devices. Another car radio system was initiated by the European Space Agency (ESA). However none of them has really become successful to date.

The main competitive system for any of the digital radio systems mentioned is audio and radio via the Web. The delivery of digital audio files via the internet became possible in the early 1990s, partly thanks to the new and efficient audio coding system of MP3 (MPEG-1 Audio Layer III). In addition, the first attempts to stream audio over the internet were made, and became more popular with the introduction of essential software like Real Audio. Streaming has made it possible to listen to audio while also receiving it, instead of first having to download a file. Web radio had come into existence and, despite the low quality, the number of radio stations offering audio streaming services grew quickly.

As the internet is the one true platform for converging media, it became the main delivery system for thousands of web-only radio operators and a central supplement for most (radio) broadcasters. The main disadvantage, however, is that – in contrast to broadcasting – the more listeners there are, the more bandwidth expense there is for the programme operator, on top of the expensive music royalties for global distribution (via the internet). Moreover, mobility or portability was until recently also an issue, but this is changing rapidly with the proliferation of broadband wireless and mobile internet access (e.g. WiFi, GPRS, 3G, 4G, 5G, etc.). Then there is time-shifted downloaded radio, like podcasting. The latter is based on software that enables people to subscribe to and download audio files in a user-friendly way directly to a computer or portable device (often in combination with Really Simple Syndication (RSS)) feeds (O'Neill et al. 2010). Finally, when limiting radio to only music there are several new (social media) initiatives for downloading and streaming music files, like Last.fm, Spotify, Pandora, Rhapsody, iHeartRadio, Google Music and iTunes Radio. The streaming audio services in particular are becoming increasingly popular (*The Economist* 2011).

CONTENT SERVICE PROVISION

Content service provision can be defined as providing a bundle of audiovisual content services to end-users via a sales, rental or ad-supported business model (Donders and Evens 2011). In digital broadcasting these audiovisual services are being offered to people via three types of distribution channels and platforms: (1) network-based audiovisual services; (2) audiovisual services via the Web; (3) and portable audiovisual services (Leurdijk et al. 2006).

NETWORK-BASED AUDIOVISUAL SERVICES

The first type refers to the network and provider-based audiovisual services, traditionally offered to end-users by a distributor or network operator. Here digital television and radio are offered on a provider-controlled and managed network via four means. First, there is the terrestrial network based on airwaves, like the traditional broadcasting network, where the audience receives the main (public or private) broadcast channels Free-To-Air (FTA) via antennas on their house or on their television set. While the analogue version has gradually been switched off in many Western countries (Van den Broeck and Pierson 2008), large areas of the Asian, Latin-America and African world continue using the traditional analogue broadcasting system (Turner and Tay 2009). Here the broadcaster is often also the distributor, and in that way has a direct link with audiences through the different TV and radio channels. In a digital broadcasting setting, however, a (new) independent (private or public) distributor can enter the game, integrating different digital terrestrial signals and offering these in packages to audiences.

Second, audiovisual services provision can come from the cable companies that offer, within their geographical service area, multiple levels or 'tiers' of TV and radio programming selections, ranging from standard broadcast stations, public access channels (available in the community), and basic and premium cable networks via coaxial cables (CATV) (Mullan 2008). For example, in the US two corporations control the vast majority of local cable systems: Time Warner Cable and Comcast. However, for a long time cable was only one-way and could not offer (the more profitable) interactive value-added services like video-on-demand (VOD), internet services and personal video recorder (PVR) programming. By coupling television with broadband internet, a return channel was created. At the same time, the digitization of the television signal spectacularly extended the capacity of the cable to carry channels (until then a serious bottleneck) (Van den Broeck et al. 2011). The CATV network operators in the analogue as well as in the digital era have always acted as intermediaries for providing content services, being able to decide which channels to transmit. Yet regulatory obligations and possibilities also play a crucial role, such as the regulation of competition (e.g. rules on Multiple-System Operators (MSOs) in the US) and obligatory transmission of certain channels (e.g. must-carry rules in the EU).

Third, the so-called Direct Broadcast Satellite (DBS) reception refers to the ownership or rental of an outdoor satellite dish that permits users to receive television programming from one of several satellites used to transmit audiovisual signals. This was one of the first networks to become digital and therefore the first digital broadcasting to experience widespread adoption. In the case of reception via satellite, people in principal have a larger offer of channels and thus have more freedom in the selection of channels. By aiming their dish at the satellite of their choice, people are able to receive the offers on that satellite, as long as it is unencrypted and freely available. However, in practice people often choose a specific package (e.g. BSkyB, Canal Digital), which means they do not readily adjust their dish to change satellite signals (if at all possible). In the US, the two main DBS players are DirecTV and DISH network.

Fourth, network-based audiovisual content reception is based on upgrading the conventional telephone lines (of twisted pair copper) to a broadband network based on xDSL technology (Digital Subscriber Line, with x standing for the different flavours of DSL, e.g. ADSL). The latter network enables the transmission of audiovisual signals via the Internet Protocol (IP). That is why this type of content provision via the traditional Public Switched Telephone Network (PSTN) is commonly indicated as IPTV. The central actors are therefore the telecom operators, like the public telecommunications organizations in the EU or any of the divested operating telephone companies in the US (i.e. Regional Bell Operating Companies or RBOCs). The revenues of these telecom players for transporting voice over the network are melting away owing to competition between the operators and the transition to the internet (e.g. VoIP (Voice over IP) services like Skype). Hence, these operators are urgently looking for new income streams, preferably based on rich data transport (e.g. digital television, HDTV) and are therefore eager to conquer the digital broadcasting market (Van den Broeck, Bauwens, and Pierson 2011).

AUDIOVISUAL SERVICES VIA THE WEB

Another type of distribution platform is the web, which is becoming increasingly popular. This is different from IPTV mentioned earlier, as we make a distinction between a managed (telecommunications) network offering audiovisual services based on the Internet Protocol (accessed mostly with a set-top-box connected to a television set) and audiovisual services on the world wide web via a broadband internet connection (accessed mostly with an internet browser on a computer screen). In the latter case the content service providers usually do not own or manage the network, which means, in principle, less control for the provider and more for the user.

The providers in this type of service provision are also called 'Over-The-Top' (OTT) players (non-infrastructure-based internet companies), using the ('unmanaged') public internet as their distribution channel, providing an aggregation service either for commercial or for user-generated content (UGC) (Bain and Company 2007: 83). There are far more providers of audiovisual services streamed on the web than ones active via the network-based services, partly as the Web has a more global scope. Given the rise of broadband internet, we see a rise in the supply and use of these types of distribution platforms, especially as video forms the largest part of the rise in internet traffic.

Case Study 3.3: Internet Television

In internet television, or so-called 'Over-The-Top' (OTT) television, a distinction is made between professional (mass) content and content by (professional) amateurs. On the professional level, the important (US) actors are Apple, Google, Netflix, Hulu and others. However, we also observe the ways that content creators and content aggregators sometimes try to skip the distributors as intermediaries, and in that way offer their audiovisual products and services directly to the consumers. In this way, public broadcasters such as the BBC (UK) and the NPO (the Netherlands) successfully offer OTT audiovisual services (respectively iPlayer and Uitzendinggemist. nl). Even content creators in other fields (like newspaper and magazine companies) deliver audiovisual services via the web. But network-based distributors are also trying to seize the opportunity and use the web as a complementary platform for distributing content, thereby creating additional economies of scope. For example, in Flanders (Belgium) this is done by the cable distributer Telenet (partly owned by the US company Liberty Global Group) with a service on the web and on portable devices (orginally marketed as 'Yelo'). As a countermove the Flemish broadcasters offer their own OTT services via the internet on different devices. They are doing this in a joint initiative (marketed as 'Stievie'), as well as separately (by offering access to their programmes on the internet). On the (professional) amateur side there are several web outlets that combine amateur and professional video, like YouTube or Vimeo. But many blogs also use video material on the net.

The digital broadcasting landscape becomes even more complex with the gradual integration of the web and television. This refers not only to connecting computer devices with television sets, via a standardized input–output system (e.g. HDMI), but also to computers that are able to receive and even record regular television (e.g. Windows Media Centers). A particularly relevant development is the introduction of so-called 'connected television' or 'hybrid television', referring to television sets with an internet connection. This opens up new ways of consuming regular television content besides internet content (whether or not in the form of 'apps').

PORTABLE AUDIOVISUAL SERVICES

Digital audiovisual services are also increasingly delivered via mobile devices (e.g. smartphones like Google Android-phones) and portable platforms (e.g. tablet computers like the Apple iPad). On the one hand, this refers to mobile applications that are closely linked to linear audiovisual programmes. Popular examples are voting via SMS texting for a song contest or music charts, and sending in short messages to live programmes. On the other hand, the evolution in portable audiovisual services also refers to an extension of the screens that can be used for consuming video and audio content.

Besides the traditional (large) screens that are fixed in one or more places in the home (large-screen television sets or desktop computer screens), people are now also able to watch live and recorded television on other (smaller) screens in different places in the home (most often via a wireless internet connection) or even outside the home in public spaces. The latter is possible through the internet (e.g. WiFi, broadband mobile networks like 3G or 4G) or – to a lesser extent – via digital broadcasting systems (e.g. DVB-H, ISDB-1seg). The additional screens in the home are sometimes also put forward as so-called 'second screens', to be used as an interactive device complementary to the main television screen (e.g. accessing enhanced features related to a show, connecting with friends through social media, giving comments using Twitter) (see Chapters 4 and 6).

CONTENT SERVICE AND TECHNOLOGY CONSUMPTION

In the end the consumption of audiovisual services at home or elsewhere, whether or not in combination with the internet and other additional applications (e.g. Electronic Programme Guide) happens through a specific user conduit (Küng et al. 1999), the end-user device and its interface. This area is subject to significant uncertainty because developments are governed by user response to the various innovative products and services that are being developed and introduced. Traditionally this was typically a television or radio set, but in the digital era many more electronic consumer devices have entered the field as main devices (e.g. large-screen LCD televisions, LED televisions, tablets) or as an enhancement for broadcasting (e.g. set-top boxes, EPGs, connected TVs, Blu-ray players, home theatre systems). The end-user stage is constantly changing as a result of technological developments and is subject to the same forces of fragmentation that are evident elsewhere in the value network. So while this role was generally only taken up by consumer electronics manufacturers (e.g. Philips, Samsung), the information technology and telecommunications companies now play an increasingly important role (e.g. Google, Apple, Microsoft). In addition, some television via internet can be consulted via game consoles (e.g. Sony PlayStation, Microsoft Xbox) or other devices (e.g. Roku). The competition for taking on this role is especially fierce since 'customer ownership', i.e. the actor who succeeds to be in direct contact with the consumer and do the billing, is perceived as crucial for holding or gaining a powerful position in the value network.

The digitization of audiovisual content opens other (legal and illegal) ways of acquiring movies, series and other programmes. The most notorious one is streaming or downloading content via peer-to-peer networks, using systems like BitTorrent or 'online social storage' (Musiani 2010). It is sometimes argued that (younger) consumers do not want to be dependent on the (sometimes inconveniently arranged) programme schedule of the content aggregators and distributors, and therefore skip the billing and take matters into their own hands. However this requires specific software and particular computer skills that ordinary television watching does not need.

This chapter has discussed how digitization is tied in with the economic and technological perspective of the broadcasting industry. The next chapter (Chapter 4) will elaborate on what digitization means for production in broadcasting, at the level of the process and at the level of the product or outcome. The level of process refers to an investigation of how digitization interrelates with the people who are responsible for the production of television and radio content. This is the professional workforce of the broadcasting industry, but also complemented by audiences becoming producers. The level of outcome discusses the changing role and meaning of content. In this way we find how the techno-economic developments are intensely linked with the social and cultural evolution of production in the digital era (Chapter 4).

Chapter Summary

■ This chapter discussed how the technological, economic and regulatory transitions that have instigated digitization and convergence are reshaping the media- and sector-specific organization of the broadcasting industry. We therefore analysed how the industry is (re)organized and how business economics are changing in relation to key technological changes in distribution, transmission and reception.

■ First, we elaborated on the basic concepts in media economics that are crucial to understand the techno-economic evolution of the digital broadcasting industry: competition, value networks and business models.

▶

■ We then applied these key concepts to frame the further discussion of the digital broadcasting industry. From a value network perspective, broadcast media in the digital era are no longer structured linearly into a value chain. The different chain stages, representing bundles of fully integrated processes, are instead being 'unbundled' into separate business roles that extend over different industrial domains (content, internet, software, telecommunication ...). These business roles are organized according to five main service-delivery phases: (1) content creation and production, (2) content aggregation and packaging, (3) content distribution, (4) content service provision and (5) content service and technology consumption.

■ In particular, the role and significance of the content service provision has changed substantially in the digital era, where the different audiovisual services are now being offered via three types of distribution channels and platforms: (1) network-based audiovisual services, (2) audiovisual services via the web and (3) portable audiovisual services.

4 PRODUCTION IN THE DIGITAL ERA

Jenkins' (2006) term 'convergence culture' is often used to grasp the post-2000 televisual culture that emerged from the turbulent dot.com bubble years in the second half of the 1990s, which led to the rise and fall of many so-called dot.com companies (Caldwell 2004). Most of these businesses, including web designers, retail websites, e-commerce, live-streaming-video websites, internet portals and online music stores, had a short life because of failing business plans, too little knowledge or concern about consumer interest, and excessive financial risk-taking. However, their visionary ideas, wild plans and creative projects unquestionably triggered the established production industries (i.e. studios, production companies, networks, public service broadcasters, commercial stations) to connect with the so-called dot.com world. The impact of this flirtation (and, in some cases, marriage) between the old, established cultural producers and the newcomers has quickly become apparent. Hence, today, audiences are engaging with TV series, news and entertainment on the internet. Media professionals are investing in web-based ancillaries of radio and television programmes. Media users have sometimes become participants in the production of media content. And programmes are 'repurposed' by the industry (Bolter and Grusin 2000; Caldwell 2008) in order to fit in all the 'spreadable media' (Jenkins, Ford, and Green 2013) or multiple media platforms, both official and unofficial, where they can circulate and be accessed, received, shaped, reframed and remixed by media users. Hence the so-called COPE-model – create once, publish everywhere (Jacobson 2009) – has become the guiding philosophy in broadcasting production (Bolin 2011; Jensen 2014: 229–30).

Convergence culture, in Jenkins' (2006) terms, is both a consumer-driven and corporate-driven process, made possible by the proliferation of portable, wearable, user-friendly, affordable (for some parts of the world) and multifunctional media. It bears the promise of a reconfiguration of media power, of a new media aesthetics and economics, where people share and exchange media commodities, come and work together around media content and engage in the endless and worldwide circulation of audiovisual texts. Although Jenkins acknowledges that these forms of grass-roots participation are probably not accessible for the entire globe, the activities shown on YouTube, in fan communities, peer-to-peer networks, etc. are all indicative of this new culture. The other side of convergence culture is the growing or consolidating power of media companies learning quickly 'how to accelerate the flow of media content across delivery channels to expand revenue opportunities, broaden markets and reinforce viewer commitments' (Jenkins 2006: 18). It is exactly this tension between actors that frames the discussion occupying many scholars in the field of media studies, and which revolves around this key question: are the changes we witness symptomatic of more fundamental ruptures in the sphere of production, or, if anything, are they to be considered as merely intensifications of the process of the industrialization, commercialization and commodification of broadcasting?

This chapter attempts to give an overview of the key trends in today's broadcasting production culture, by considering them in light of technological innovations and economic developments. Two particular discussions will guide this chapter. The first considers the relationship between media professionals or cultural producers, on the one hand, and media users, still also known as audiences (see Chapter 6), on the other hand. Some argue, with Jenkins (op. cit.) as one of the main exponents of this idea, that this relationship is challenged and reconfigured in the age of digital culture, with ample consequences for the forms of audiovisual culture and concrete media texts (see also, for example, Curtin 2009). In particular, the audience's techno-logically based ability to interactively engage with media, its technological and media knowledge (hence the tech-savvy and media-savvy audience), and, as a consequence, the growing interest of media professionals in fans' ideas, interpretations and readings of media texts (TV series, reality games, etc.) are marking the era of a new ecology of production and consumption. In making use of terms that express the blurring of production and consumption, like 'produsage' (Bruns 2008), 'prosumer' (Toffler 1980), 'intercreativity' (Berners-Lee 1999) and 'user generated media', this point of view is exemplified. Others, often from an economic perspective, refute this analysis and draw attention to the structural characteristics of the economic and industrial system that broadcasting production is part of and, in Bourdieu's (1983) terms, to the culturally and socially shaped fields of institutionalized, established and legitimized practices of both media professionals and media audiences (see, for example, Deuze 2007; Hermes 2013).

The second discussion deals with the newness and innovativeness of contem-porary broadcasting culture (see, for example, Dawson 2007). Digital culture and web-based technologies are believed to engage innovative transformations in media production as a means to facilitate professional activities in, for example, the sphere of news and fiction production. When it comes to broadcasting output, television studies point at the upsurge of new media texts and new engagements with old media texts (Jenkins 2006). Last, digitization makes the process of customizing audiovisual content, modes of address, but also advertising to the particular tastes, preferences and interests of various types of niche audiences, fan communities and subcultures technologically feasible and, more importantly, economically profitable (Dovey 2008; Evens et al. 2010). In particular, Anderson's Long Tail theory argues that in the age of online retailers a substantial fraction of revenue is generated from consumers' demand for digital niche content (the so-called tail sales), and that the 'future of business is selling less of more' (Anderson 2006). Evidence shows that consumers still prefer the big cultural hits and superstars (Webster and Ksiazek 2012), but it goes without saying that niche markets, catering for the specialist, and sometimes eccentric, tastes of consumers, have widened the traditional logic of the audiovisual mass marketplace (Anderson 2006).

THE FIELD OF MEDIA PRODUCTION

Traditionally, the process of broadcasting production consists of four distinct stages, in which (1) programme ideas and their potential realization are devised (development); (2) after budgets have been defined, the concrete preparations for production are planned (pre-production); (3) the actual shooting takes place (production); and (4) the editing of the shooting material is completed (post-production) (Bignell 2004). In each stage, many media workers are involved, often salaried but equally well freelance (Hesmondhalgh and Baker 2011), all taking up different roles and working practices that cover the creative, technical, organizational, supportive and managerial aspects of making radio and television programmes, both frontstage and backstage (Ursell 2006).

Although both radio and television production entail a wide range of media professionals, there are great differences between radio and television production in terms of cost (radio is much cheaper), time (on TV, time is much more scarce), complexity (crews for television programmes are much bigger) and flexibility (radio programmes and items can be easily dropped as the situation demands) (McLeish 2005), which makes television a business that is much more controlled, stringent and risk-conscious. Indeed, studies show that tensions and conflicts are an intrinsic part of the television production process, where script-writers and directors have to defend and negotiate their ideas, and hence their creative autonomy, within the broader managerial framework and mission that TV executives envisage (Gitlin 1994; Manning and Sydow 2007). On the other hand, radio producers and presenters also have to meet the overall station format, the expectations of the well-defined audience and the programme schedules and music playlists station managers have determined (Hendy 2000), a development that in the light of the extensive branding of radio stations has become common practice (see Chapter 5).

Because of the increase of broadcasting hours and the new platforms of consumer supply and distribution, which have led to growing competition and increasingly insecure production revenues, the sector of broadcasting employment has moved from a permanently employed to an increasingly freelance, casual and flexible staff, exposed to intensified exploitation, short contracts, insecure wage agreements and struggles over professionalism (Deuze 2007; Ursell 2006). A significant part of the production of television and radio programmes takes place within the broadcasting companies, for example in the newsrooms or in production units within the television channels and radio stations specifically set up for fiction production. Alongside this, a whole industry of both large-scale (predominantly and long-established in the US) and small-scale independent production companies and studios has grown significantly. They produce on a project-by-project basis, sometimes in co-production with channels and networks, a range of TV series, quiz game shows, reality game shows, documentaries, etc., or develop factual and fictional television formats that regularly become worldwide TV franchises (see 'The TV franchise *Big Brother*' case study below), shown in national variants on many screens all over the world.

Case Study 4.1: The TV Franchise *Big Brother*

One of the most notorious European examples of a worldwide TV franchise (which is the licensing of the intellectual properties of a TV programme format) is *Big Brother*, the 'factually-based' entertainment show that was created by the Dutch TV producer John de Mol (Endemol) in 1997, first shown on Dutch television in 1999 and is still being aired on many television stations worldwide (Esser, 2010). This programme format has become emblematic for convergence television, as it combined television broadcasts and live internet feeds on the programme's webcam site. And already in its first edition audiences were invited to interact directly with the participants and the show by accessing the website's online chatrooms and participating in the audience vote (Turner 2006). It is also an iconic example of the worldwide success of the reality TV genre (see section 'Something New?'), and more particularly the proliferation of so-called day-to-day surveillance TV shows (Andrejevic 2002). What we are watching is the non-scripted private social interaction among 'ordinary' people, obviously framed, edited and steered in particular, extraordinary directions in order to make the interaction more emotionally loaded or discordant. For instance, in *Big Brother* we watch the participants doing everyday intimate activities, like sleeping, eating, washing, but this all takes place in an artificial setting and under controlled circumstances. Interestingly, these reality TV participants often become famous people and get a lot of attention in gossip magazines and the tabloid press, like 'traditional' TV celebrities (i.e. professional actors, performers, singers, etc.) (Holmes 2004). Their celebrity status, however, is attained not by their professional talent or hard work, but by their so-called ordinariness, degree of truth and authenticity, which audiences are carefully scrutinising when they watch the show and the participants themselves are increasingly aware of (Andrejevic 2002; Holmes 2004; Turner 2006). As such, *Big Brother* has played an important role in the shifting relationship between professionalism and amateurism in TV production and economy.

PROFESSIONALISM AND AMATEURISM

Considered from a media-sociological point of view, and inspired by Bourdieu's field theory (see Hesmondhalgh 2006; Hesmondhalgh and Baker 2011; Manning and Sydow 2007), broadcasting professionals make up an interesting social field. According to Bourdieu's thinking, the social field is characterized by 'struggles over positions, which often take the form of battle between established producers, institutions and styles, and heretical newcomers' (Hesmondhalgh 2006: 216). Creative freedom and innovation are only possible when the field and its structuring practices enable them (op. cit.: 216). Hence, the way different broadcasting professionals are positioned and related to each other within this field can be understood in terms of the social practices they enact day to day on the ground, in the companies they are working within, in the broader professional networks of which they are a part. Within this specific field one can discern categories of perception about what practices are rewarded or circumscribed, or what practices are un/thinkable and un/doable (Hesmondhalgh 2006: 216). For example, directing a quality TV series is estimated to be higher in professional value than directing a soap opera. Presenting a reality game show is valued lower than hosting your own talk show.

Hence, following Bourdieu's ideas, the technological opportunities that digitization has brought about, and the social and cultural practices that go together with these technological developments, will not necessarily reconfigure the field of broadcasting production, since the field itself, i.e. its agents, their practices and the way they perceive and evaluate their and others' practices, is still guided by strong ideas about professionalism (Bignell 2004: 154–5) and artistic and editorial integrity (Deuze 2007). Observations of today's interplay between producers working inside the media industry and those outside show signs that the traditional distinction between professionalism and amateurism, which involves not only the field of broadcasting but all spheres of cultural and symbolic production, is both challenged and reproduced. On the one hand, both the pursuit of profit and creative talent makes the broadcasting industry and its professionals encourage and embrace the creative labour and participation of fans and amateurs. On the other hand, for fear of losing artistic integrity and revenues (which boils down to the constant concern about intellectual property rights that 'protect as well as limit the range of possible markets and applications of cultural products' (Deuze 2007: 96)), media companies and workers are also responding adversely to people's reworking and reusing of television and radio programmes (Jenkins 2006: 37). And whereas the availability of inexpensive high-quality computer editing systems have given low-budget students and independent producers many opportunities to make programmes that can stand the comparison with professional productions, both audiences and commissioning broadcasters have much higher expectations for the quality of those products because of advances in post-production technology (Bignell 2004: 152).

Reflection: Professionalism and Amateurism

Think about the media-made material (e.g. photos, videos, pictures, vlogs, animations), if any, you have created yourself and uploaded to the internet. Would you consider it to be the equivalent of the professionally made content found on the internet? What do you think are the most important reasons to differentiate between professional- and amateur-made content: the maker's training, talent, equipment, audience reach, creativity, social network or social status?

In addition to these factors, the relationships between media professionals and media audiences are embedded in long-standing views about the power of media and the 'aura' surrounding media creativity (see Couldry 2000), demonstrated for example in the way media companies, radio and TV studios, stage sets or newsrooms are often hard to enter, not only physically speaking (having difficulties entering the building or the set without permission), but also professionally – as employment in the media sector is predominantly built on network relationships within relatively closed communities of media workers (Deuze 2007; Manning and Sydow 2007; Ursell 2006). What is more, the media industry itself (although it increasingly solicits creative input from fans), plays an important role in perpetuating the cult of the professional media worker by providing so-called behind-the-scenes material on the internet and DVDs, such as interviews with cast and crew, documentary shootings of film and TV sets, etc. (Caldwell 2008). Likewise, talent scouting among non-professional media-makers, both on the internet and in myriad TV programmes, has become an important aspect of media-industrial practices, whereby creative and enterprising amateurs and fans are quickly incorporated in the industrial processes of media production (Caldwell 2006; van Dijck 2007b). However, indicators of the success of new programmes still rely heavily on sources and moral orders coming from the field of media workers themsevlves (Bignell 2004; Ursell 2006). Clearly, these established ideas about the media as bastions of power and creativity reflect upon people working in the media. Media professionals, and television and radio programme makers in particular, have been traditionally credited with great technical (e.g. camera work, montage, music, sound dubbing), creative (e.g. script, animation, DJ) and communicative skills (e.g. interviewing, hosting, talking), which the audiences (albeit often criticizing the media output) do not necessarily ascribe to themselves. Although many people want to work within the media industries, jobs are scarce and people who manage to enter the field of media work prefer to define their work in terms of pleasure, self-expression, self-enterprise and self-actualization

(Deuze 2007; Ursell 2006: 161). As Bignell (2004: 155) states, 'the television industry is a self-enclosed world, with powerful internal hierarchies and codes of shared knowledge, status, competition and gossip'.

The availability and usability of an ever-expanding range of not-so-expensive, high-quality technologies, with simple instructions, have affected the closed field of broadcasting professionals. Some argue that radio and television workers have evolved from skilled specialists to skilled generalists, with people other than the traditionally film-trained cameramen having acquired shooting and editing skills (Ursell 2006: 150). This process of role convergence (cf. Huang et al. 2006: 93), where media workers are combining *different* specialist forms of work that used to be diversified, invites re-skilled and multi-skilled professionals. Indeed, the fast-paced spread of technology and ongoing digitization in media work result in a structural demand for highly skilled specialists, and have put pressure on older media practitioners (Deuze 2007: 93–4, 183). This is, for example, keenly felt among news professionals, who are increasingly producing news content for multiple media platforms on a routine basis, requiring sophisticated technological skills (such as computer-assisted reporting and multimedia production), for which neither media employers nor schools are providing appropriate training (Huang et al. 2006; see also Deuze 2007).

FAN- AND AMATEUR-BASED PRODUCTION

In addition to professionally based transformations, the field of broadcasting production has been challenged and confronted by new types of material and new forms of aesthetic expression that creative people, from amateurs to potential film- and television programme makers, are uploading and sharing on the internet. Indeed, the web 2.0 world and television's convergence with it is often being seen – and now and then even celebrated – as one that is obliterating the consumer-producer binary in the cultural sphere (Deuze 2007; Gillespie 2012; Jenkins 2006). Also known as user-generated content (UGC) and user-generated media (UGM), and created outside the field of professional routines and practices, content is created by people – formerly known as audiences, publics or consumers – using the internet, their smartphones, their tablets and their game-consoles not only to consume, but also to produce, share and distribute cultural texts (Shao 2009). Although a large part of this kind of everyday cultural consumption, production and distribution is not aimed at securing economic profit, it has consequences for the professional and industrial field of broadcasting (Bolin 2011: 26; Deuze 2007). Interestingly, young independent film-makers and production companies set up their own websites and are making extensive and intensive use of YouTube to show, promote and share their productions with potential audiences, and likewise catch the attention of the media industry. But even their small-scale productions cost money, and if they want to survive and get noticed, they have to look for other channels to have their productions financed, such as crowd-funding (see, for example, Wong Fu Productions in the US, wongfuproductions.com).

One of the most conspicuous early examples of how a non-professional media-maker succeeded in galvanizing attention of a mass audience was the 19-year-old Northern-American student Jennifer Ringley, who in 1996 started registering the routine and intimate details of her daily life via a webcam in her college dorm room. Ringley maintained her JenniCam website, with more than five million hits a day (see Andrejevic 2004: 61), for more than seven years and emerged an internet personality. But perhaps more importantly, the JenniCam website heralded the coming of a TV production culture that comes close to what non-professionals can do themselves with compact and cheap film-making equipment and without a script, costumes, actors, television sets, etc. The consolidation of reality as a professional form of entertainment (as explored further below) was the undeniable corollary of the webcam burst (Andrejevic 2004), and is today rampant on YouTube, where videoblogging or vlogging has become a dominant form of user-generated content (Burgess and Green 2009b).

But also in many other respects, the expanding group of amateurs creating, filming, recording, uploading and sharing audiovisual material has become more visible in professional production contexts. First, news programmes publish significant levels of user-made content on their ancillary websites, or even integrate camcorder footage – also known as 'found footage' (Bignell 2004: 125) – in their televisual newscasts. For example, television news items on natural disasters, traffic accidents, fires and explosions, but also war coverage, regularly mix professional camera footage with amateur recordings – often of poorer quality. This practice is again not new, as it was, for example, already used in television's coverage of the assassination of the US president John F. Kennedy (see Zelizer 1993). The practice is, however, increasing, and the amount of material about the murder of the president made available today on the internet is incomparable to what television in the 1960s was showing. However, to give user-generated content a place in news magazines and on news websites does not necessarily mean that media corporations are leaving the storytelling to others – rather they are incorporating these stories-from-below inasmuch as they 'bolster their own position as leading social storytellers' (Couldry 2012: 24).

Perhaps more interesting are the various forms of so-called 'grassroots appropriation' of media (Jenkins 2006: 281). Here we are dealing with the transformation of media content, a process that has become 'increasingly central to how popular culture operates' (ibid.: 282), and is no longer the exclusive territory of small niches of extreme fans. Ordinary radio listeners and television viewers are ripping, reformatting, uploading, downloading and accumulating audio and video in new, mostly web-based environments (Kompare 2010: 81). All kinds of fans, finding each other in global fan communities of actively engaged people, are remixing material made available by the television industry into self-made creations. Although this kind of work is often described in terms of democratization and empowerment, critical scholars ask pertinent questions about the creative labour viewers and listeners are performing for the media industry. In becoming enthusiastic producers of content, audiences are also becoming cheap producers of value, who all contribute happily to the profit of the media companies (Hayward 2013). Reality TV formats (see section 'Something New?'), are a good example of this: rather than letting paid actors and writers do the work, the ordinary people in the TV shows and on the websites are creating the economic value of these programmes (Andrejevic 2004).

THE INDUSTRY FIGHTS BACK

The fit between people's growing engagement with radio and TV programmes (thanks to the technological developments in the fields of internet, second screen and mobile technologies), and the industry's search for new profitable projects, have incited broadcasters and producers to experiment more systematically with so-called transmedia storytelling, a new production process that has been enabled by digitization and media integration. In transmedia storytelling, one story or narrative is multiplied across various technological platforms, all creating their own, distinct story-worlds (Bolin 2011: 97; Jenkins 2006: 21). Hence: a TV series is discussed on a fan's forum; the life, personality and background of the characters is being elaborated on the programme's website; unresolved scenes are shown and talked about on social media; alternate reality games are being played with the main characters. This trend in audiovisual production is in essence dual. On the one hand, it is driven by the market and non-artistic motivations of media companies engaged in several media sectors all trying to grasp the benefits of associated intellectual property rights (which Jenkins (2006) terms 'corporate convergence'). On the other hand, it is also made possible through the creative and non-market-motivated input of media users, as it opens up opportunities to enthusiasts for sharing with other devotees, both amateurs and professionals, the interpretative work they are performing when they are watching television ('grassroots convergence' in Jenkins' (2006) terms) (Bolin 2011: 98). This duality raises questions about the aesthetic innovation versus formulaic repetition transmedia storytelling is bringing about (Jenkins 2006). Since transmedia storytelling guarantees more intense forms of participation and engagement, even after the series episode has been watched (Caldwell 2004; Örnebring 2007a: 457), there is no doubt that the so heralded interactivity made possible in transmedia storytelling is industry-driven (see section 'Broadcasting and interactivity').

Case Study 4.2: Alternate Reality Games

Alternate reality games are a form of internet-based mystery game in which participants are immersed in a fictional world based on a TV series, and engage in the collaborative process of unravelling the mystery. For example, the ABC-TV (US) series *Alias* was accompanied by two alternate reality games, one launched by the network ABC and one produced by fans. Örnebring's (2007a) study concludes that despite the differences found between industry-produced and fan-produced alternate reality games, both types of alternate reality games contribute to the corporate goals of marketing and brand-building, as well as to fans' pleasurable interaction with fictional story-worlds.

In general, the broadcasting industry and broadcasters are more aware than ever of the importance of good stories (Fusco and Perrotta 2008: 95). In the branding of channels, fiction has become a crucial tool to control the market or seek a market, and to articulate a network's or station's brand identity. Although various old-established television and radio content and genres seem to survive the transition towards digital broadcasting – for example, soaps are still being broadcast daily on many television channels – the pressures of convergence (that is, the demand for multi-platform content, immediacy, hypermediacy and new forms of serialization (Fusco and Perrotta 2008: 91–3)) result in the unremitting exploration and exploitation of new ways of bringing storytelling to people. However, experiment and innovation remain relatively marginal and small-scale because of a high sensitivity to business risks in the world of television (see Ursell 2000, among others), and an increasing insecurity about audiences' expectations from, loyalty to and interest in broadcasting as it used to be (Fusco and Perrotta 2008).

SOMETHING OLD?

Clearly the convergence between the 'old' mass media mode and the 'new' interactive mode of media production, between the push and pull media production modes, produces if not ruptures, then indisputably ripples. Many argue that digitization radically alters the conditions of culture, as it dissolves the boundary between production and consumption (Bolin 2011; Poster 2005). Although traditional institutions, conventional practices and established structures are not overthrown by new technologies and are still not obsolete, we are indeed witnessing transitions. Interestingly, new media forms are, in Bolter and Grusin's (2000: 15) terms, honouring, rivaling and revising old media forms, and as such contributing to the phenomenon of remediation – identified by Bolter and Grusin (2000) as the defining characteristic of the digital media age. 'What is new about new media comes from the particular ways in which they refashion older media and the way in which older media refashion themselves to answer the challenges of new media' (Bolter and Grusin 2000: 15). The coming of new media entails a process of give-and-take, of mutual influence and shaping between new and old media aesthetics, appearances, visuals and styles (Cooke 2005). Hence, rather than arguing that new media are casting everything overboard, ample evidence can be found that digital media are remediating their predecessors. Traditional broadcasting media content is borrowed, reused or, to employ a term often used within the contemporary culture industry itself, repurposed (Bolter and Grusin 2000: 45; Caldwell 2004).

And so we can see that today's broadcasting industry shows particular interest in relatively old content that fits in well with the economy and culture of convergence. In particular, content that has the potential of provoking 'strong audience engagement and investment', that enables 'multiple entry points in the consumption process' (via a wide array of transmission channels, delivery technologies and tie-ins), and that allows media users to 'more quickly locate new manifestations of a popular narrative' is given priority (Jenkins 2003: 284). This is not necessarily new content, as the prevalence of sport in convergent radio and TV culture demonstrates. Indeed, sport seems to take the lead in the move from analogue to digital broadcasting, as it combines two key principles of the television industry: it is relatively inexpensive to produce and guarantees big market shares (Johnson 2009; Kruse 2009). Major media conglomerates in the world, such as media mogul Rupert Murdoch's News Corp. (Horrocks 2004: 57; Kruse 2009), but also national broadcasting companies (broadcast stations and cable channels) make significant efforts to purchase the rights of sports games such as rugby, football and baseball, and use this type of content as a 'battering ram' to enter new markets (an expression coined by Rupert Murdoch – see Kruse 2009: 182). Sports content also links up with digital broadcasting's interest in betting – which the emergence of horse race wagering via TV and dedicated horse racing channels clearly demonstrates (Kruse 2009). Remarkable also is the relatively old-fashioned style of the coverage of sports such as baseball in the US (Bolter and Grusin 2000), and soccer, tennis and cycle racing all over the world (although instant replays and flash statistics on the screen can definitely be considered the influence of the logic of hypermediacy that new media have brought with them).

Music also has proven itself a very rewarding form of multiplatform content. In answer to the pressing questions about television's creative impasse and future, music, one of the oldest forms of entertainment, has been rediscovered as a good basic ingredient for convergence culture (Fusco and Perotta 2008; Holmes 2004). This is shown in reality programme formats like *The X Factor* (SYCOtv, UK) and *Idol* (FremantleMedia, UK) or *The Voice* (John de Mol, the Netherlands) – and their many local variants worldwide. In these shows unknown people with an extraordinary musical talent are confronted by an expert jury panel with at least one stern and therefore famously controversial jury member (see, for example, Simon Cowell in the

UK, and the intertextual references made to him in the movie *Shrek 2*). Exactly this mixture makes up an extremely successful cocktail for audience entertainment and participation. If media industry professionals are constantly looking for the water-cooler TV topic, these formats are definitely a bullseye (Holmes 2004; Ytreberg 2009).

SOMETHING NEW?

One cannot ignore the shaping role of technology in the aesthetics of today's digital broadcasting output. An important development here is the emergence of the so-called 'camcorder cult' (Dovey 2004: 557) – obviously much older than the dot.com burst, but which paved the way for what Ouellette and Murray (2009) have called the 'celebration of the real' and 'the entertaining real', and Andrejevic (2004) has named 'the push for the real' and 'the promise of the real'. The rapid and cumulative development of 'camcorder' technologies (from the home video camera in the 1980s to the webcam in the 1990s and the camera phones of the 2000s) has indeed had a significant impact on the form, content and look of many contemporary films, documentaries and television series (van Dijck 2008). Clearly, the quest for reality was inscribed in the broadcasting media from the very start (Ouellette and Murray 2009: 4; Uricchio 2004a). There had been some concrete historical precedents – such as *An American Family*, the first real docu-soap aired in 1973 on PBS (US) (Andrejevic 2004; Ouellette and Murray 2009). However, it was not until the 1980s that home video recording became established, with Sony's release of the first camcorder for non-professional users. TV stations then started broadcasting programmes that consisted of home video footage delivered by ordinary people ('amateurs'), mainly showing funny or spectacular scenes from everyday life, accompanied by a professional voice-over.

These early forms of user-generated content marked the coming of a popular, low-budget mainstream television genre (also known as reality TV), containing a wide spectrum of subgenres (from TV reality crime programmes and court programmes, reality sitcoms and docusoaps, to reality game shows such as gamedocs, dating programmes and talent contests), all prominently aired. The latter have become particularly successful among wide and diverse audiences. Although the diversity of these subgenres is substantial, they all share a set of characteristics that have been intensified by the digitization of broadcasting and today's internet culture. First, they all show a strong tendency towards the recognition of private lives and intimate moments (Dovey 2004: 557; van Dijck 2008: 71). Second, ordinary people (i.e. non-actors or non-celebrities) very often play a leading role in these programmes. Third, reality TV genres are hard to group into one genre or one category. They are 'hybrid', often combining and fusing elements from different TV genres and registers that used to be strictly separated from each other – for example soap (a fictional genre) and documentary (a factual genre) (Ouellette and Murray 2009).

Finally, they all hold a strong claim to reality, meaning they are spontaneous, unscripted or lightly scripted, or at least create the illusion that what the audience sees on television is how it 'really' happened when the camera registered it. These characteristics foster the aura of 'authenticity' around these programmes, as the scenes and people shown in them are perceived as 'genuine' rather than fictitious. Some have argued that audiences are often very well aware of the constructed nature of reality TV programmes, and find enjoyment in playing with the 'real' and 'unreal' ingredients of the programme. Nevertheless, the 'authenticity' claim of reality TV is one of the main explanations for its popularity among audiences globally, as it is believed to connect strongly with the overall search for the 'true' self and the genuine life that characterizes postmodern consumer culture (Ouellette and Murray 2009; Rose and Wood 2005). Hence, even the appearance of showbiz, rich or celebrity people and their extra-ordinary lifestyles in reality TV programmes such as docusoaps (*Keeping Up with the Kardashians*, aired on the US channel E!), and reality sitcoms (*The Osbournes*, premiered on MTV, *Astrid in Wonderland*, aired on the Belgian channel VIJFtv), is also framed by the question of how special these people 'really' are in their daily existence, when they too show remarkably ordinary routines, practices and concerns (Kompare 2009).

This so-called non- or minimally scripted TV entertainment was perhaps initiated via 'the camcorder cult', but it was the particular political-economic context of the 1980s, with its growing competition among broadcasters and networks (see Chapters 2, 3 and 5) that made it flourish (Andrejevic 2004: 90). Searching for new revenues, instant TV hits and low-cost productions, reality entertainment programmes met the needs of a very competitive television industry that was constantly in search of steady viewers, as well as hit programmes with low production costs. Hence, the worldwide omnipresence of reality TV genres in today's broadcasting schedules does not come as a surprise, but rather fits extremely well with the economic and technological changes and insecurities broadcasting is now going through. Since the broadcast content of reality TV is heavily built on the work and contributions of unpaid amateurs, reality TV has much lower production costs than, for example, TV fiction, which in combination with its broad popular appeal has made it a very profitable form of content for the industry. What is more, 'the real' and 'reality' are a type of content that, unlike fiction, floats more easily between different media platforms, such as television and the internet (Edwards 2012).

It may also be more flexibly scheduled – unlike TV fiction there are no real 'seasons' in reality TV – and is therefore beneficial for broadcasters that have to respond quickly to declining audience shares and viewing figures – another high risk for today's broadcasting industry. This flexibility also dovetails with digital broadcasting's non-linearity, i.e. the erosion of television as flow, and the emergence of on-demand and online viewing practices that challenge television's traditional sources of revenues and rhythms (see also Chapters 3 and 6) (Bennett 2011; Chamberlain 2010). First, reality TV offers new opportunities to integrate advertising into the programmes themselves, which has become a crucial quest for both broadcasters and advertisers, as people can more easily skip ads (Andrejevic 2004: 90; Ouellette and Murray 2009: 2). Second, although it is still customary for broadcast companies to announce the new television season's programmes, the extensive development of audiovisual content on the internet and on DVD has created a television culture that is no longer based on eagerly looking forward to new programmes (which characterized the age of scarcity – see Chapter 2), but is rather oriented towards plugging as quickly as possible into 'must see' TV shows. Reality game shows such as *Big Brother* (see box 'The TV franchise *Big Brother*'), *Idol* (created by Simon Fuller, FremantleMedia, UK), *The X Factor* (created by Simon Cowell, SYCOtv, UK) and *Survivor* (created by Charlie Parsons, UK) are all good examples of what Scannell (2002) has called 'event TV', in which 'the live and living moment of the event itself' (op. cit.: 272) is crucial to its popular appeal (see also Turner 2006). Reality game shows are all conceived within a larger 'ontology of expectations', building 'momentums from week to week' (Scannell 2002: 273), i.e. a form of serialization, sometimes more than nine weeks long, creating significant sociability as large groups of people are discussing the content both offline and online (see also Ytreberg 2009).

Particularly with the advent of the internet and interactive talk-back technologies, these reality TV genres have proved the true 'killer applications' of media convergence (cf. Jenkins 2006). This is because they enhance audience participation, emotional engagement and investment – which is, for both advertisers and broadcasters, crucial in terms of market reach, consumer attention and viewer loyalty (Bennett 2011; Bolin 2011; Fusco and Perotta 2008; Gripsrud 2010c; Ytreberg 2009). The content of these broadcasts is very often far from new – talent scouting shows were already shown on television in the 1950s (see, for example, Bauwens 2007) – but the degree of involvement that is made possible today through social-communicative technologies, such as Twitter, has never been seen before.

With digital broadcasting the features that have been ascribed to postmodern television have clearly intensified. The playful mixture of genres, the wealth of intertextual references and allusions to different forms of popular culture (like music and film), and the increasing reflexivity about the medium itself, its conventions, and its traditions are all characteristic for today's television (Bignell 2004). Indeed, in the search for new audiences – or, as McLuhan argued, consequential to electronic media culture – so-called 'genre instability' (Kompare 2010: 79) or 'hybridization' (McLuhan [1964] 1994) has become one of the key characteristics of contemporary television culture (Fusco and Perrotta 2008: 89; Spigel 2004: 4;). This is demonstrated in content where the boundaries of genres, fact and fiction are blurred – in so-called reality spoof shows (satirizing reality TV's conventions through sophisticated fictional productions – Ouellette and Murray 2009: 5), or TV news parody shows (challenging the social authority of news – Druick 2009), for example. Overall, fiction as a major category of television content has not lost its importance, but on the contrary has crystallized into a wealth of TV series that often intersect various genres and narrative formulas (Spigel 2004).

Case Study 4.3: *The Blair Witch Project*

Perhaps one of the first iconic examples of the novel fictional aesthetics that convergence culture has brought about is the US low-budget pseudo-documentary horror film *The Blair Witch Project* (1999) by Daniel Myrick and Eduardo Sánchez, which became a significant box office hit (see Schreier 2004). In every respect the film plays on the audience's fascination for improbable and suspenseful stories. In combining different genres – i.e. documentary, horror film, and video diary – the film clearly transcends the traditional boundaries between fact and fiction. It tells the story of the disappearance of three young film students, who as part of a class assignment, plan to spend three days and nights in the woods to find the legendary Blair Witch, and make a documentary about their quest. They vanish, no trace can be found, except for the material they filmed during their trip. *The Blair Witch Project* was shot entirely via video camera (imitating non-professional camera work), and the fact that the directors and actors were unknown to the wider public increased the film's claim to reality and authenticity. The film was heavily promoted through an intensive viral marketing strategy on the internet, providing ancillary content about, among other things, the three students' lives and the legend of the Blair Witch. The film thereby posed as a documentary, made by real people and about existing events, and did this quite successfully – as a significant part of the audience was at least temporarily uncertain about the reality status of the film (Schreier 2004). *The Blair Witch Project* is therefore a paradigmatic forerunner of the internet's contemporary role in promoting fiction conceived for television and film screens.

All television programmes are being rethought of as convergent television content – Caldwell (2004: 49) points out that in the rhetoric of the media industry and media workers, 'content' has replaced 'programme'. They are conceived as a type of 'snack television' that can be navigated and consumed via micro-fragments – such as key scenes, news items, or interview extracts – which are continuously reproduced and distributed on multiple platforms without loss of technical quality (Bolin 2011: 12; Fusco and Perotta 2008: 94). This is demonstrated well in the evolution of TV newscasts that have an increasingly 'scannable' style of information presentation, and which have visually converged with new media design (such as ticker-tape delivery style, eye-popping visuals and updates panels). This convergence in the appearance of news media produces an environment in which news items can easily float among different platforms because they share the same 'information module structure' (Cooke 2005: 41). Transmedia storytelling also demonstrates the ways in which TV products are 'conceptualized as a series of networked texts' (Gillan 2011: 2) that prompt fans to track the content across multiple media platforms.

But likewise, old, iconic and quality-labelled TV series are launched as season DVD box sets (e.g. *The Sopranos* (HBO, US), *The Simpsons* (Fox, US), *The Wonder Years* (ABC, US)), or are rerun via platforms such as Netflix, enabling television audiences not only to re-enact nostalgic television memories and fetishize their fandom (Kompare 2006, 2010), but also to consume television 'without the "noise" and limitations of the institution of television' (Kompare 2006: 352). The importance of TV fiction is abundantly clear in the DVD industry. Unlike other cultural products that have become less bound to their tangible carriers (e.g. music, newspapers, books) with the advent of digitization (Bolin 2011: 25), television programmes have become increasingly materialized. That is, the introduction of DVD technology in 1997 created a culture of television consumption built upon a concrete form (Kompare 2006). Likewise, online access to television content through retailers like Apple's iTunes store, video players/sites enabled by the media industry or illicit peer-to-peer networks demonstrates that the old television art of telling stories in serialized form is surviving the digital culture of YouTube.

BROADCASTING AND INTERACTIVITY

One of the key promises that emerged in early experiments with interactive TV (see Bignell 2004: 265–7; Caldwell 2004: 53), and has from the very start figured in popular discourses of digital broadcasting (Van den Broeck et al. 2011), is the dream of user–user, user–producer and user–content interactivity via mass media. Whereas radio was initially conceived as an interactive medium (see Chapter 2), and has always been more open to listeners' input into its programmes (such as into call-in programmes), both radio and television broadcasting have for a long time been understood as monologue forms of mediatized quasi-communication, where real interaction between the context of production and the context of reception was not made possible (Thompson 1995).

Although the broadcasting industry likes to stress the interactive opportunities of digital broadcasting, and 'interactivity' has become a significant feature of media marketing, critical questions are often being raised about the innovative features of today's interactivity, and its prospects for consumer empowerment and democratization (see, for example, Gripsrud 2010c: 16–18; Nightingale and Dwyer 2006). The opportunities for playfulness, choice, connectedness, information collection and reciprocal communication (cf. Ha and Chan-Olmsted 2004: 624) might have increased as an integrated feature of television and radio programming. Yet influential interactivity – that is, the creative and active intervention in the development and narrative of radio and television shows – is still very limited (Gripsrud 2010c: 17), and what is more, not necessarily desired by large segments of the audience (Bird 2011; Simons 2014). Obviously, participating in programmes such as quiz shows and reality TV formats, voting on candidates or commenting on the topics, guests, participants and hosts of particular programmes via Twitter, Facebook and the programme websites, are all forms of audience interactivity that have increased in the age of convergence culture. However, viewers choosing their own camera angle, influencing the narrative flow in their favourite series and/or deciding which test the participants in the reality contest game must undergo remain largely excluded practices.

From an industry and technology perspective, experiments in what has been called 'interactive narration TV', 'collaborative TV' (see Bachmayer et al. 2010) and 'participation drama' (Bolin 2011: 102) have been regarded as largely successful. Owing to non-online TV's technical limitations, it is mainly internet- and mobile-based technologies that can offer viewers the possibility of altering the narrative flow of a programme. For example *The Truth about Marika*, a drama series broadcast by the Swedish public broadcaster (SVT), combined fictional TV content with web and mobile services, in search of Marika's lost friend. This participation drama has often been praised for its truly 'mixed story world' as the plot narration took place in the physical world (via mobile technology) and on the internet, as well as on TV (Bachmayer, Lugmayr, and Kotsis 2010). Other broadcasters have also set up interesting experiments with this kind of interactive narrative TV, for example ITV (UK) with the soap opera *Emmerdale* (Bignell 2013: 292), where viewers could choose the ending with a vote. Another example is the telenovella *Emma*, broadcast on the Flemish public service broadcaster (VRT), that persuaded viewers to solve a murder case in the series. These forms of TV fiction are new insofar as they encourage viewer collaboration across multiple platforms. None of these series, however, achieved high ratings – not even the iEmmy-awarded series *The Truth about Marika* (Bolin 2011: 101) – and in addition, all had a very short lifespan.

TV fiction that involves viewers in constructing or deciding upon storylines is rarely realized (exceptions are NBC's (US) *Heroes* and Channel 4's (UK) *City of Vice* (Bachmayer, Lugmayr, and Kotsis 2010)), largely because the creation of multiple simultaneous linear plot lines is extremely expensive and complicated (Bignell 2004). It is also because of a lack of interest on the part of the audiences, who still want to be immersed in a narrative told and delivered by professional storytellers (Simons 2014). Precisely this immersive power of good stories has, however, proved to be one of the 'killer applications' for digital broadcasting. In particular, the phenomenon of transmedia storytelling clearly demonstrates that television programmes are 'often at the top of the hierarchy when it comes to multiplatform productions' (Bolin 2011: 106).

That is to say, narratives as developed in TV series and reality TV formats remain the core of the broadcasting industry's success, whereas the activities that viewers and fans display through mobile and web services remain predominantly informational (receiving news about the show through RSS feeds, accessing additional audio and video streaming, consulting the specialized Wiki about the TV series) and communicational (connecting with other enthusiasts via social networks, forums and blogs) (Bachmayer, Lugmayr, and Kotsis 2010).

Interactive interventions in fictional TV narratives on the part of the audience may also not be common practice owing to increased competition within the TV market (see Chapters 3 and 5). Since the 1990s, we have seen the development of more and more TV series with a narrative structure that encourage so-called 'fannish' explorations and readings, on multiple media platforms, and provide more fuel to activate a fan community (as above) (Örnebring 2007a: 451). Many of these series also described as a new kind of 'cult' or 'quality' television – examples of which are found not only in the US (e.g. *Lost*, originally aired on ABC; *Alias*, also on ABC), but also in smaller countries such as Belgium (*De Parelvissers*, aired on the Flemish public broadcasting channel Eén), and Denmark (*The Killing*, aired on the public broadcasting channel DR1) – deliberately create gaps and obscurities in their complicated storylines so as to stir viewers' imaginative engagements (cf. Örnebring 2007b; Sconce 2006).

Until now, therefore, the most successful forms of 'offline' TV interactivity to have a real impact on broadcasting content are so-called game- and competition-based features of interactivity – that is, tele-voting. These features are currently the only profitable and thus applied uses of the loudly heralded coming of interactive television (Bachmayer et al. 2010; Bignell 2004: 267–9). The benefits of other features of interactivity between the viewer and the TV industry, i.e. fan-, information- and programming-based, 'are still more a marketer's wish than a consumer's reality' (Ha and Chan-Olmsted 2004: 623).

Chapter Summary

■ This chapter began with Jenkins' notion of convergence culture to understand the changes within the field of broadcasting production. Building on this framework, we argued that the developments and trends in how and what audiovisual content is made, are both consumer- and corporate-driven. We introduced two matters that are often raised for discussion in contemporary literature. The first deals with the shifting relationship between media professionals and audiences, the latter often believed to be more able to participate in and interact with media offers. The second matter addresses the forms and types of today's broadcasting output and their alleged newness.

■ In order to examine the mutual relations between media producers/workers and consumers/amateurs, we concentrated further on the social characteristics of the field of media production. Furthermore, we explained how ideas about 'media work' are part of larger, long-established views of culture, creativity, power and aura.

■ We next moved to forms of fan- and amateur-based output, and explored how the media industries are responding to these new types of creativity. Several authors critically brought to our attention that the big media companies and established production industries are successfully incorporating and capitalizing on the creative work of ordinary people.

■ The second half of the chapter reviewed forms and genres of audiovisual content that are revitalized (such as music and sports) and intensified (such as reality TV) in the digital age of broadcasting. In particular, reality TV and its authenticity claim, next to the primacy of good TV fiction, has become one of the key constituents of the digital broadcasting offer.

■ We concluded the chapter by critically exploring the alleged promise of interactive television formats.

5 CHANNELS IN THE DIGITAL BROADCASTING ERA

Traditionally considered as the intermediary players between the media-professionals and producers (discussed in the previous chapter) and the audiences and consumers (tackled in the next chapter), this chapter addresses the shifting meaning of television and radio channels in the digital era. The coming of new content-delivery platforms on the internet questions the traditional distributor and exposer-role that channels and stations have played ever since the start of broadcasting. From an industrial perspective these channels are labelled as 'content aggregation and packaging' (see Chapter 3). For a long time, channels were the exclusive intermediary between, on the one hand, the producers and creative people who make the programmes and who were until the 1980s mostly staff members of the broadcasting institutions (cf. Bignell 2004: 138) and, on the other hand, the audience. New players, with Hulu (US) and the BBC's iPlayer (UK) as the outstanding examples of the paradigmatic shift in delivering TV content, have seriously challenged this exclusive go-between position. With the proliferation of digital broadcasting channels, amateurs and fans creating, uploading and sharing their own audiovisual material on the internet, and new providers aggregating and recombining media content, the traditional intermediary role of broadcast channels is being challenged. Whereas channels have always played, and to a considerable extent still do, an important role in controlling access to audiovisual programmes and content and managing the radio and TV experience of the audience, it is clear that the radio and TV experience has been dislocated from the scheduled flow (Bennett 2011: 2). Online video services, such as YouTube, Hulu, the BBC's iPlayer, music streaming services such as Last.fm and Spotify, and multiple, mostly mobile delivery devices, i.e. mobile phones, iPods, iPads, games consoles, are disrupting the old balances in the broadcasting ecology. How are the broadcasting institutions and companies dealing with this, and how does this affect the role and position that broadcasting channels play in an increasingly digitized media environment?

FROM BROADCASTING TO EGOCASTING

It is hard to discuss the changing meaning and role of TV channels and radio stations without using the term broadcasting. Broadcasting refers to 'a cultural form where audio-visual material is disseminated [...] in a continuous, sequential form – a flow – from some central unit to a varying number of anonymous people who receive the same material at the same time – or, [...] roughly the same time' (Gripsrud 2010c: 9). Even the narrowcasting channels that began to proliferate in the 1980s in their free-to-air and subscription variants meet this definition. Although these so-called 'concept channels' (Bolin 2011: 91) transmit material that is much more restricted and aimed at much narrower niches of specialist audiences and distinct communities of taste, age and interest, they often have a mass or worldwide reach (e.g. CNN, MTV, National Geographic, Al Jazeera and Nickelodeon).

However, the features that are considered essential to the notion of broadcasting and the traditional modus operandi of TV channels and radio stations have been progressively challenged by a wide and growing gamut of delivery platforms, services and devices. This chapter examines the phenomenology of the broadcasting experience, i.e. channel experience, which is built upon the 'televisual essence' of 'flow' (Uricchio 2004b; Williams [1974] 2003), modernity's interest in the 'simultaneity' (Moores 1995; Uricchio 2004a) of mass reception and the 'sense of liveness' (see, among others, Bourdon 2000; Couldry 2004a; Moores 1995) that are gradually being reconfigured.

Obviously, this reconfiguration has a long history. It begun with the remote control in the early 1950s, which marked the beginning of the era of the personalization of television technology, and continues with the VCR, which eroded channel control over what audiences could watch and when they could tune in (Rosen 2004; Uricchio 2004). With the remote control viewers could easily switch between TV channels and look out for enjoyable programmes without leaving the couch. The VCR brought about even greater audience freedom. This technology enabled people to watch consecutively TV programmes that were often broadcast at the same time on different channels (i.e. counter-programming strategy) or to enlarge the choice on offer by watching, for example, films previously recorded or hired in video shops. The Sony Walkman (1979) and, more than two decades later, the iPod (2001) put radio stations to the test in their role as music providers (Berry 2006). Today, the multitude of on-demand services, pocket-sized devices and smart TV platforms, developed by the game, entertainment, internet and electronics industries (see Chapter 3), exemplify the nexus between thorough individualization and accelerating technologization in the Western industrialized parts of the world and mark the coming of a new era. Some have called this the age of 'egocasting' with its deprecatory connotation of the personalized, narrow and hence egocentric pursuit of audiovisual material that meets one's personal taste (Rosen 2004). Others have denominated it the era of 'programming your own channel' (Rizzo 2007) with its more optimistic undertone of consumer sovereignty and democratization, i.e. throwing off the yoke of schedule-driven and routine-oriented broadcasting by creating your own playlists. Although, in particular with respect to TV, one can say that the content is often transmitted and received as broadcasts or narrowcasts, coming from well-known established channels, which are firmly rooted in national and international communities (Gripsrud 2010c: 9), it goes without saying that the era of customization shakes the ecological system of TV channels and radio stations to its foundations. The younger generations of media users especially are abandoning radio stations (Albarran et al. 2007), redefining the notion of channel loyalty by combining multiple music platforms in their everyday music consumption patterns. Or else are simply unable to identify what broadcasting institutions, radio channels or TV networks are (Lotz 2007: 2).

FROM CHANNEL SCARCITY TO CHANNEL ABUNDANCE

Looking back upon the history of radio and TV, the digital advances that are affecting the meaning, identity and role of channels for media users and society, are part of longer-term developments within the channel landscapes (see Chapter 2). These trends are culture-specific and connected to economic, political and geographical differences. Hence, there are differences between the Western world and the other continents of the world, between Europe and the US and within Europe between the old liberal democratic countries and the former communist countries. However, the move from semi-single-channel environments to multi-channel environments, undermining centralized broadcasting arrangements, has been a general trend all over the world, starting with the increasing volume of radio stations in the 1970s to the subsequent proliferation of TV channels in the mid-1980s (Ellis 2000; Moran 2005; Terzis 2007). In Europe, for example, the public service broadcasters (in Western Europe) and the state-controlled broadcasting companies (in former communist Eastern Europe) dominated the channel landscape (Bolin 2011: 10). And unless more liberal governments in these countries authorized commercial channels or the geographical location of the receivers permitted them to access more easily foreign TV channels and radio stations (e.g. the cross-border spillover of neighbouring country's channels in Europe (see Mills 1985)), these central broadcasters had a monopoly. But even in the US, the oft-quoted model of a liberal radio and TV market, it was only from the mid-1980s onwards that cable and subscription channels started to erode the dominance of the three big TV networks, NBC, CBS and ABC, and the American viewers quickly began to explore multi-channel environments (Lotz 2007: 12). This move from the 'era of scarcity' (1950s–1970s) to the 'era of availability' (1980s–1990s) (Ellis 2000), or, viewed from a US perspective, the shift from the 'network era' to the 'multichannel transition' (Lotz 2007), took place in changing political-economic contexts (i.e. liberalization, deregulation and commercialization of the broadcasting systems) and went along with important technological developments both in the transmission of radio and TV signals (cable, satellite, CD) and in actual radio and TV devices (remote control, VCR) (see also Chapter 3).

Although all these shifts reconfigured the 'age of plenty' (Ellis 2000) or 'post-network era' (Lotz 2007), it is the twenty-first century that marks the real coming of what we could call channel abundance, and this is happening all over the world. In China, for example, the number of channels available for Beijing's households increased spectacularly from 13 in 1997 to 48 in 2004 (Yuan and Webster 2006). In the UK, between 1992 and 2009 the number of households receiving television via cable or satellite grew from 2.3 million to 13 million. Digital terrestrial television (DTT) increased from nearly 800,000 in 2002 to 14.8 million users in 2009 (Golding 2010). The Indian channel environment exploded to 350 TV channels (Sonwalkar 2008: 124). The transmission of audio and video by digitally compressed signals has facilitated the launch of countless digital-born channels that take up much less space than analogue ones, and so make room for far more channels. A lot of these channels, which are run and owned by multinational American-based corporate broadcasters (e.g. News Corp., Viacom Inc., The Walt Disney Company, Time Warner), public broadcasters (e.g. Australian Broadcasting Corporation), local commercial media groups and amateur radio and TV organizations, are conceived according to the model of narrowcasting and aim for a wide range of specialist niche audiences, from children (e.g. ABC3 in Australia; Disney Junior), women (TLC in the UK) and men (Dave in the UK and Ireland), via sports enthusiasts (Motors TV in the UK; Juventus Channel in Italy) to lesbian, gay, bisexual and transgender communities (Logo TV – developed by Viacom and MTV Networks). Many of the new specialty channels have a cross-border and pan-continental reach and try to gather diaspora audiences, migrant communities, expats and larger cultural regions (e.g. Indian language-channels such as Bollywood Times in Canada; The Asian Food Channel (Sonwalkar 2008: 124)). However, not all new channels are successful in attracting audiences, advertising or funding, and the continuity and life expectancy of many of these channels remain rather precarious.

Next to these digital channels, whose modus operandi is still based on traditional ideas about exhibiting and distributing audiovisual content, the age of plenty has brought the arrival of new so-called post-linear providers that, in aggregating, packaging, distributing and delivering traditional radio content (music in particular), and television programmes (TV series, documentaries) on the internet, are bypassing the established radio and TV stations. Commercial music services on the internet, such as Spotify (launched in Sweden in 2008) and Last.fm (founded in 2002 in the UK), distribute music in the form of streamed content and explore new business models in their pursuit of profit (Bolin 2011: 10). Although these online services cannot be considered as broadcasting channels in the strict sense of the word, as distribution tools they probably encroach more deeply on the traditional ontology and interface of TV channels and radio stations that is still heavily built on flow, liveness and linearity. Gateways such as TiVo, Hulu, YouTube, Google TV and Apple's iTunes suite represent what might be called new 'metadata-based aesthetics'. They are interfaces, software-described visual intermediaries between media users and content, organized along the ideas of customization, navigation and consumer-control (Chamberlain 2010). Clearly in the postlinear TV landscape, channels are increasingly approaching their audiences through visual interfaces that resemble the online gateways to audio-visual content. For example, on most of the European TV-channels in-vision announcers, which gave the experience of watching TV a feeling of liveness and flow, have taken a backseat (see Van den Bulck and Enli 2014). Likewise, broadcasting channels have taken up the customization discourse in their on-demand and interactive services both on air and on the web. And major broadcasters, like ABC, NBC and BBC, have quickly realized that 'official' web streaming sites where people can watch recent episodes of new shows and full seasons for selected classic shows, are not only a good promotion tool, but also an attractive alternative for the often lower-quality online video services offered by YouTube (Kim 2012: 60). The US online video distributor Hulu provides a good example of this trend. Initially only telecom partners were involved (AOL, MSN, Facebook, Comcast, Myspace, Yahoo!) in this joint venture, when it was founded in 2007 and launched for public access in 2008 as a television streaming service. Today, however, three big US audiovisual cultural industries (NBC Universal Television Group, Fox Broadcasting Company and Disney–ABC Television Group) own the enterprise. Last, video-sharing websites, of which YouTube, launched in December 2005, is the most notable example, have enabled the set-up of a wide variety of both free-to-view and subscription channels, on which both amateurs and media corporations can upload, view and share videos (see Case Study 5.1 'YouTube channels').

Case Study 5.1: YouTube Channels

Most of the YouTube channels are personality-driven or dedicated to particular personalities, and show videos in which persons want to share their talents with the world or in which famous persons, like politicians and ▶

pop stars, appear (e.g. Obama-channel). In particular humoristic YouTube channels are abundant and very appealing to young audiences. A good illustration is Rémi Gaillard's YouTube channel. Gaillard is a massively popular self-declared humorist from France, sharing practical jokes and pranks with more than 2.6 million subscribers. Another example is Higa's YouTube channel, set up by the Northern-American actor, comedian and producer Ryan Higa, with more than 8.3 million subscribers. YouTube also accommodates thematic channels that are dedicated to a wide variety of topics ranging from hard information (e.g. critical investigative journalism, politics, education) to soft information (e.g. fashion, makeup, pets, sport).

After Google's purchase of YouTube in October 2006, we also find an increasing number of examples of YouTube channels that are set up by mainstream and institutionalized cultural industries and media corporations using YouTube branded channels as just another distribution tool for promoting their programmes and connecting with their audiences and fan communities, for example through 'webisodes' (a 3-5-minute episode of TV shows for web showing only). Although these types of channels are compromised in YouTube's amateur-led culture of user-generated content, as the protest against Oprah Winfrey's YouTube Channel in November 2007 illustrated, they equally show that broadcasting materials are increasingly disembedded from TV (Burgess and Green 2009b; Kim 2012; Wasko and Erickson 2009). Interestingly, even though the concept of YouTube channels is different from that of traditional broadcasting channels as they are built on discontinuous and on-demand user experiences, channels on YouTube nevertheless create a horizon of expectations for their potential audiences. For example, they let their users know how many videos they upload on a weekly basis, and as such they are constructing a feeling of continuity.

SURVIVAL OF THE FITTEST: BRANDING

As shown worldwide, the growth of delivery platforms through which media users can access audiovisual material thoroughly impacts on the control the old broadcasting channels used to hold, if not as the monopoly then surely as the oligopoly, over the viewing and listening diet of people. Although the interest of the public in these channels has not dwindled as fast as some people forecast (Tay and Turner 2010), all over the world the trend of fragmentation has been visible for quite some years now (e.g. Webster 2005). In the UK, to give but one example, between 1992 and 2009, the share of people watching BBC1 and ITV1 has been steadily dropping from 37 per cent and 44 per cent respectively in 1990 to 22 per cent and 18 per cent in 2008 (Golding 2010: 214). Smaller TV channels, like US local variants, are also experiencing declining viewership (Greer and Ferguson 2011).

Because of the proliferation of new channels and platforms, the growing competition for audience share, the fragmentation of audiences and the increasing sovereignty of consumers to create their own media experiences and compose their own media menus, branding has become a key issue for all media industries, and not least for radio and TV stations (Johnson 2012). For media consumers as well, the age of offer abundance has sharpened the importance of brands as the originator of products and services that one can put trust in and rely on, and that are compatible with their needs and expectations (Andjelic 2008; Caldwell 2008; Chan-Olmsted and Kim 2001; Grainge 2010; Gray 2010; Turpeinen 2003). Hence, the process of channel branding boils down to the creation of a 'holistic identity to viewers and consumers' (Caldwell 2008: 245). In building a stable set of expectations, core values and promises of a particular, distinct, consistent and recognizable content and style, in which audiences put their faith when they tune in, channel branding strategies are built on emotional affinity and relationship (see, among others, Born 2005; Caldwell 2008; Grainge 2010; Van den Bulck and Enli 2014).

Radio broadcasters were aware of the importance of branding themselves in a distinct way much earlier than TV broadcasters. Even by the 1960s, radio stations, especially in the US, were entering the age of plenty, with multiple stations offering whole day-long or even twenty-four hours of programmes with comfortable sound quality and multiple delivery technologies producing distinct listening habits (e.g. on the transistor radio and car radio). This crowded radio environment, together with the pressure from television, urged listeners and producers to look for new and distinct programme formats and soundscapes (Cordeiro 2012: 494–5; Nyre 2008: 134). Radio broadcasters quickly recognized that branding a channel identity through niche programming and distinctive, original and recognizable station sounds were a means to get noticed amid the abundance of media offers. Not only commercial, but also public service broadcasters have made significant efforts to identify the core values of their radio stations and manage them as a communicable identity. By 1967 the BBC had already reorganized its radio output into four channels, each serving a particular part of the audience, catering for distinct music tastes and creating a feeling of belonging to particular communities of interest (see Cordeiro 2012; Grainge 2010: 53).

Reflection: Radio Channels and Audiences

Can you think of two examples of current commercial or public service broadcast radio channels, and decide who their audiences are and how they attempt to reach them through particular programmes?

In the early 1990s, as the offspring of increased competition in a 'noisy media marketplace' (cf. Chan-Olmsted and Kim 2001: 75), television broadcasters joined in the trend of channel branding. In search of a high brand familiarity, perceived quality and positive associations that are linked to the brand – the resources of what is called 'brand equity' (see Aaker 1991; Chan-Olmsted and Kim 2001) – the new TV channels, e.g. US cable television and European commercial television channels, explored and applied different strategies to build a strong brand. They invested heavily in original programming, in trailers, in logo designs or so-called 'channel idents' (Meech 2001) – the graphic and animated symbols representing channels (see Bignell 2004: 16) – and in brand slogans and awards. But by the end of the 1990s public television broadcasters, eager to establish themselves in a liberalized TV market, also came to recognize the necessity of a strong brand identity built upon values like quality, reliability and solidity (Biltereyst 2004). Through a whole repertoire of marketing (logos, promos, videoclips), production (design news studios, brand personalities, modes of audience address, in-vision announcers, jingles) and programming strategies, TV channels have tried to build an identity, differentiate themselves from others and build a loyal relationship with audiences (see, among others, Born 2005; Caldwell 2004; Chan-Olmsted and Kim 2001; Gray 2010; Hoynes 2003; Meech 2001; Van den Bulck and Enli 2014).

Reflection: Branding Tools

Can you give any examples of the branding tools listed above: marketing (logos, promos, videoclips), production (design news studios, brand personalities, modes of audience address, in-vision announcers, jingles) and programming strategies?

This trend has only intensified with the digital convergence of the twenty-first century. In particular, the rise of broadband platforms, upon which a large offer of TV-content is aggregated and made available to audiences en masse, challenges both old and new channels to re-invent and rebrand their so-called 'channel identities' or 'channel formats' (Bolin 2011: 90) continuously. Rather than broadcasting, 'brandcasting' (Grainge 2010: 45–6) has become the paradigm in which all radio and TV channels are involved today. The demands of a highly competitive, technologically fast-moving multiplatform environment and the shifting culture of media use puts pressure on all channels to 'project their brand identity in more dynamic and tactile ways' (ibid.: 46). This entails a massive investment in promotion, sometimes at the expense of advertising space, to remind viewers and listeners of the channel to which they are tuned. Equally, cross-media promotion has become a well-tried strategy (Caldwell 2008; Greer and Ferguson 2011). A good example is the extension of the channel's brand identity on official internet sites devoted to major shows and TV series. Here TV channels aim to enhance the passive, linear box-experience by offering fan-, game-, information- and programming-based features built on playfulness, choice, connectedness, information collection and reciprocal communication. All these forms of interactivity are believed to be crucial to sustain viewer loyalty (Ha and Chan-Olmsted 2004), or, to put it in marketing terms: they contribute to a form of emotional bonding, i.e. bonding through aspirational values ('We understand each other, we share the same values, the same spirit' (Kapferer 2012: 140)).

Case Study 5.2: Interactive Branding Strategies

In some cases, the degree of viewer engagement on TV show websites is intense, as for instance with the websites dedicated to competition entertainment shows like *The X Factor* (Jerslev 2010) and *Britain's Got Talent* (Enli 2009). Not only these reality TV genres (see also Chapter 4), but also TV series, such as *Supergirl* in China (Yang and Bao 2012), are heavily built on viewer engagement and participation through interactive websites, extra information on the episodes, actors, characters and participants, back stories, behind-the-scenes clips, interviews, discussion forums. Even social network systems like Facebook and Twitter are being used to build a connection with audiences (Greer and Ferguson 2011). As such, TV channels create their own 'water cooler spaces' (McIntosh 2008: 74).

Newer branding strategies, enabled through media convergence, also include those built on transmedia storytelling, a concept introduced by Jenkins (2006) (see Chapter 4). In coordinating several technological platforms, online and mobile, a greater narrative whole, i.e. a series of different multiplatform connections with the original narrative, is constructed. Motivated by market rationales and fan participation, this new form of telling (and selling) TV stories and TV experiences is used by channels to stand out from the crowded and intransient media landscape and to connect briefly but intensively with their consumers and fans. Both scripted (like the Fox TV series *24*) and non-scripted television programme formats (e.g. *Big Brother, Idol* (see Chapter 4)) have proved that TV stories and the channels that are broadcasting them are often at the top of the hierarchy when it comes to multiplatform productions, and that they can create a large crossover market (Bolin 2011; Scolari 2009).

Another online strategy used by the established broadcasters is to reorganize and re-profile their audiovisual content so that it fits the 'structure of feeling' (Williams, [1961] 2001) of the YouTube generations. One of the many examples here is the internet channel of the Danish public service broadcaster, Piracy TV, which is aimed at young people and remixes broadcast segments in a YouTube-wise way (Bondebjerg 2010: 122). But in other countries as well, both public service and commercial broadcasters try to establish and maintain strong relationships with their young audiences by recasting their broadcast content in dynamic, interactive, game-, event- and community-oriented online environments. In particular with respect to these audience groups, brandcasting often brings about, in Lash and Lury's terminology (2007), an extensive 'thingification' of the channel identity – and more particularly of the programmes they are broadcasting – through material objects (bags, T-shirts, caps, etc.) or 'mediation' of the brand through other media (CDs, books, blogs, events, etc.). All TV and radio stations, both commercial and public service, are applying these merchandising strategies to an increasing degree in an attempt to secure the relationship with their actual and potential audiences.

Last but not least, the audiovisual content itself (i.e. programmes, music genres, radio formats) and the way in which they are scheduled shape the format logic of channels. The increased pressure on broadcasters to compete with other distributors and exposers of TV and radio material has strengthened the importance of scheduling, hence constructing well-defined audiences as commodities for advertisers (Bolin 2011). Even in the nonlinear television era, generalist channels were making big efforts to cultivate a flow between distinct programmes, by making use of voice-overs, graphics and videoclips (Van den Bulck and Enli 2014), and 'create a symbolic universe endowed with meaning' (Scolari 2009: 599).

Case Study 5.3: 'It's Not TV. It's HBO.'

Television channels like the US premium cable and satellite network HBO (launched in 1972, and today owned by Time Warner) demonstrate that fiction has become an imperative medium to build a brand identity ('fiction as brand' (Scolari 2009: 599)) and make 'an interpretative contract' in which it proposes a world of values of which consumers agree (or not) to become part (op.cit: 599). With TV series such as *Sex and the City*, *The Sopranos*, *True Blood* and *Boardwalk Empire*, HBO is a good illustration of this branding strategy. In producing and broadcasting original, challenging and complex TV series that require various forms of media and cultural literacy, and which consequently received many Emmy and Golden Globe nominations and awards, HBO has rebranded itself as a network that is not ordinary and cheap, but sophisticated and worth paying for. As such, HBO 'has become synonymous with quality in the contemporary television landscape' (McCabe and Akass 2008: 84), communicated by its network slogan 'It's Not TV. It's HBO.' (1996–2009) or 'It's More Than You Imagined. It's HBO.' (2009–present). HBO-series express distinct values and propose 'an aesthetic, a series of textures, colors, materials, and styles that create a difference with respect to other brands' (Scolari 2009: 600).

FROM GRAZING TO BROWSING

Following Chamberlain (2011), it can be argued that the emergence of new gateways and interfaces for accessing audiovisual content is reconfiguring the way in which people engage with audiovisual content. More particularly the sense of place that has accompanied radio and TV reception for more than 50 years is making room for a placeless ontology of radio and TV. The new media interfaces (i.e. metadata-described interfaces such as TiVo, iTunes, YouTube, etc.) that can be accessed through multiple devices are reshaping television into 'non-places'; the connection to content that we want to see or hear occurs on 'transient, functional, crucial screens' (Chamberlain 2011: 233). Hence, the experience of listening to the radio and watching TV is no longer organized around the channel of distribution, but other criteria, such as genre, viewer rating and actors, are becoming guiding principles in the way people engage with audiovisual content.

Clearly, this has important phenomenological consequences for how people experience watching TV and listening to the radio. Programmes are increasingly consumed as disembedded bits and snacks, i.e. uprooted from the broader programming structure, culture and philosophy channels adhere to and that used to shape the particular 'horizon of expectations' from which people accessed audiovisual content. This theoretical concept, originally introduced by Jauss (1982), has been used in reception studies to explain how media users draw upon a structure, rooted in their time, in history and in their culture, to engage with media texts, i.e. programmes, formats and genres. This structure of predispositions is derived by former media-textual experiences. Hence, media users become familiar with particular conventions, features and rules, and when they encounter new texts, they will approach them from a set of dispositions, the so-called horizon of expectations, that they rely on (Wilson 2009: 70). Radio and TV channels have put a good deal of effort into creating a predictable and consistent environment where audiences more or less know what they can expect from the genres, aesthetic styles, personalities, music and formats on offer. For example, some stations hold a so-called highbrow programming culture and broadcast TV series, comedies and documentaries that show a more unconventional, alternative, serious, sophisticated, intellectual and aestheticized style of narrative and production. The American premium cable and satellite television network HBO is a good example of such a channel. Others endorse a more popular, mainstream philosophy of programming and fill the programming hours with genres such as comedy, action and police series and reality TV. Still others, such as for example European public service stations and generalist commercial broadcasters, aim for a general audience. Hence, they build in different time slots aimed at different age groups, like a morning and afternoon slot for elderly people and a family slot in the early evening.

Channels are flows of programming, driven by strategies to blend different programmes into a textual whole (Uricchio 2004b; Williams [1974] 2003) that stands for a particular channel experience. Drawing an analogy with Chamberlain's analysis mentioned above, channels can also be described as 'scripted spaces'; they are designed environments that invite particular reactions from their visitors and create a predestinated ontological realm of reception (Chamberlain 2011: 239–43). Precisely this experience of being somewhere, in a place, where one can be part of what is happening, in the flow of the moment, has defined for many years the phenomenology of radio and TV. Even with the coming of the multi-channel era, the availability of cable service, the video recorder and the remote control, where people gradually but steadily had the opportunity to switch between more channels, to move more easily among various programme forms and, hence, to subvert the programming strategies of broadcasters (Uricchio 2004b), the channel flow environment remained the dominant horizon of experience and expectations.

Whereas in most programme guides, both paper and electronic, channels remain an important starting point to organize viewing choices, many argue that the new media interfaces mark the disappearance of flow and broadcasting, and the coming of interface, hyperlinks and database structure (Bennett 2011: 1; Chamberlain 2011). Others agree that this is the most fundamental transformation of the user–medium relationship, but that flow will remain the organizing principle of TV. This is evident in the way in which producers, advertisers and programmers are responding to the increased programme availability and intensified control over the media offers by digging up time-tested programming techniques (Uricchio 2004b: 247).

Yet, there are also signs that the role of the channels and stations as phenomenological organizers of radio and TV reception is not completely played out. 'Choice fatigue', which is tied in with chronic 'time famine' in industrialized parts of the world (Ellis 2000: 169–71), is believed to be an important psychological side effect of the explosion of offers and services enabled by digital compression technologies (Chamberlain 2010: 86). Worldwide, from the US, via Europe to China, it makes people turn to a delineated grazing zone, a so-called 'channel repertoire' (Heeter et al. 1988), in which they move among a limited number of channels (Gripsrud 2010c: 21; Rosenstein and Grant 1997; Webster and Phalen 1997: 133; Yuan and Webster 2006). Although it has been suggested that the entire concept of repertoire might have to be reconsidered in the light of the new delivery systems (Yuan and Webster 2006: 535), channels are still a place where people make anchorage in the infinite sea of audiovisual content, and they are believed to be useful in overcoming the general problem of the 'findability' of relevant and likeable content. In our media ecology of abundant and easily accessible information and entertainment, the need for orientation and guidance on the part of large groups of media users is apparent (Turpeinen 2003). Although viewers can choose from hundreds of channels on their digital TV box, causing the audience's overall viewing time to become more fragmented than in the past, in Europe quite a number of old, first-generation TV stations continue to attract substantial shares of the public. In some Western- and Northern-European countries the public television channels gather at least 30 per cent of the viewing time (e.g. Belgium (Dutch-speaking community), Denmark, Finland, France, Italy, Poland, the UK). But the old generalist commercial channels also stand up relatively well, in spite of the proliferation of commercial TV stations (e.g. Finland, Italy) (see http://mavise.obs.coe.int). Music radio stations too, notwithstanding the impact of the iPod, iTunes, online music streaming websites and so-called homegrown YouTube musical artists, still hold a dominant position in relation to the record industry, with a concrete influence upon signing policies, release schedules, playlists, charts and the sound of popular music recordings (Klein 2009; Percival 2011).

Reflection: Old and New Media

To what extent do you still watch and listen to 'older' radio and TV channels, and to what extent has your media consumption moved to newer ones? Why do you have this pattern of media consumption in the light of issues discussed in this chapter?

Hence, the evidence shows that what we could call the old radio and TV stations are still important players in the market, and that most of the people in the world engage with audiovisual content through broadcasting networks (Simons 2013; Turner 2011). This observation holds especially when one takes a close look at the less affluent parts of the world, where the distribution and adoption of new digital TV platforms and accessories is far less universal than in the richer, mostly Western nations of the world (Straubhaar, 2007). 'Some have a choice but others do not', Turner (2011: 33) soberly notes. And all the opportunities that the new interfaces are offering to 'establish new televisual flows, finding, sorting, tagging, storing, and playing back television content across a range of media environments' is still the preserve of 'privileged viewers' (Chamberlain 2010: 86). There are states, with large publics, like China, that simply restrict access to new delivery technologies. Others, such as Australia, have a reasonable and increasing degree of media convergence, but do not have as many services and devices as the US market at their disposal (Turner 2011: 31–2). One Nielsen report on video viewership among Australians aged 16-plus estimates that 95 per cent of viewing is on the traditional television set. Television reaches 62 per cent of people in prime time, whereas online video viewing accounts for only 14 per cent. Compared to watching video on computer, which amounts to 47 hours on a monthly basis, watching TV in the home in its broadcast form rules with 99 hours, as watching playback TV – PVR-penetration in Australian households is 49 per cent – and streaming video on mobile phone amount to, respectively, 7 and 3 hours per month (Nielsen 2012: 5).

Especially in the affluent, democratic parts of the world, where the free-market TV system knows little political intervention, the reason for still holding on to old-fashioned modes of watching TV is often merely users' disinterest in and/or reserve towards innovative TV. For example, in Finland, owing to the popularity of cable television satellite, IPTV packages have difficulty in breaking through (see http://mavise.obs.coe.int). Hence, large numbers of people are just sufficiently satisfied with broadcast television and do not feel the need or want to pay for the extra services that bear the marketing promise of emancipation from the channel regimes (Boddy 2002). Even in the US, which is a leader and trendsetter when it comes to media convergence and television innovation (Turner 2011), most people 'continue to experience television as a broadcast medium, characterized by the real-time reception of linearly scheduled programme flow on standard-definition analog sets' (Dawson 2010: 96). With only 38 per cent of Northern-American households having access to video-on-demand services through their cable-providers, and a smaller share actually making use of these services, statistics show that viewers still spend ten times more hours watching television through channel surfing than via time-shifting, online and mobile television platforms (ibid.: 96–7). Hence, although the transition from analogue to digital modes of finding a way into the programmes on offer is definitely apparent, 'anachronism and in-betweenness', as Dawson (2010: 97) argues, 'remain defining characteristics of many viewers' everyday experiences of television, as they have been for the majority of television's history'. Whether this can be ascribed to socio-economic inequality is still under-examined (Dawson 2010), given that broadband internet services are far less available in the rural areas of the US, where older, less-educated, lower-income groups are more common than in urban areas (LaRose et al. 2011: 92).

PUBLIC AND COMMUNITY

The blueprint of broadcasting and, more specifically, the media logic of TV channels and radio stations, has been built upon remarkably consistent and enduring ideas of domesticity, liveness, national identity, simultaneity, publicness and sharedness (Boddy 2003; Dayan 2009; Gripsrud 2010c; Scannell 1989). All these ideas have underpinned the social role of both radio stations and TV channels, in their public service as well as commercial broadcasting forms, in modern and late-modern societies all over the world. Although the early vision of live transmissions of sound and moving images from public, mostly national, sometimes world, events to people's homes (Gripsrud 2010c: 9) is eroded by the advent of on-demand technologies of interruption and digital devices that produce highly individualized mediascapes (Boddy 2003), liveness, simultaineity and sharedness still remain a fundamental part of the specificity of channels and stations. Whereas on TV very few programmes today are really live, they at least create the experience of liveness as a 'possibility, not always accomplished but at least virtually present' (Bourdon 2000: 531), an experience that 'reminds us that it links us live to something, to a specific place ("live from") to a specific person ("live with")' (ibid.: 532). Perhaps a better term to designate this experience is what Ellis (2007: 154) has called 'sense of currency'. In using direct address, by means of TV and radio, the broadcasters (newsreaders, talk-show hosts, quizmasters, in-vision programme announcers, weathermen), by referring to the current time ('now', 'here', 'in a moment', 'today') and recent happenings ('yesterday', 'a few hours ago') in traffic information, news bulletins, entertainment programmes, sporting events, etc., 'explicitly claim to belong to the same historical moment that their audiences are living' (ibid.).

As such, liveness and currency are difficult to dissociate from shared viewing and listening (Bourdon 2000: 534; Dayan 2009: 22). Many have argued that modernity's search for simultaneity and social cohesion have paved the way for radio and TV as pre-eminently the media of simultaneous transmission, enabling a 'sense of proximity and contiguity' (Uricchio 2004a: 135), so wanted in a world that was facing disintegration in the wake of industrialism, urbanization, migration and the two World Wars, and dealing with populations that were losing their sense of place and affinity with larger communities (Moores 1993; Williams [1974] 2003). In constructing an experience of 'instant live connectivity' (Scannell 2009: 226) the broadcasting media have been an instrument for managing collective attention to news, events and popular culture and for creating 'a common factuality' (Dayan 2009: 25) or 'a daily consciousness of being a Nation' (Boddy 2003). Consequentially, within the history of Western liberal democracies, channels and stations have been considered crucial to the building of a public sphere, where shared, collective attention results in a virtual community, having something to think and talk about with others who are part of this community (Scannell 1989).

The communicative structure of broadcasting media expresses and embodies the 'we-ness' that channels and stations are shaping when they are engaging with their audiences. Hence, the phenomenological experience of watching TV channels and listening to radio stations, of consuming audiovisual content within a channel-flow environment, is shaped by what Scannell (2000) has called 'for-anyone-as-someone structures'. Watching TV and listening to the radio in such an environment is 'always, at one and the same time, for me and for anyone' (op. cit: 9). Turning to the news is an outstanding example of this double-sided sense 'that I am spoken to while knowing that millions of others are watching at exactly the same time and seeing and hearing exactly the same things' (op. cit.: 11). This individual experience is virtually a shareable experience, one that is talkable-about and that creates the possibility of 'a public, shared and sociable world-in-common between human beings' (op. cit.: 12). Even watching fiction live, i.e. as it is transmitted, may create a feeling of being part of a specific interpretive community or a national audience (Bourdon 2000: 550). Clearly, these traditional ideas on which channels and stations have built themselves for decades are eroding in the new digital broadcasting ecology. What is more, the embedding and appropriation of new technologies of simultaneity, such as the internet, in people's everyday lives are turning TV's and radio's connectivity to the simultaneous into an engagement with the recorded past. Television and radio are becoming 'a very different sort of time machine – one which permits instant access to random points' (Uricchio 2004a: 134) in recent history that have been broadcast on radio and TV.

Critical scholars and cultural commentators look with Argus' eyes at the proliferation of delivery platforms and gateways, and the resulting splintering of the audience. The disappearance of a 'shared common media culture which might form the foundation for a discursive sharing, which is in its turn the currency and core of a societally integrated experience' (Golding 2010: 214; see also Dayan 2009) is considered a serious risk to democracy and the spirit of community. Are the highly individualized mediascapes, our individualized schedules of daily life, our private mobile auditory worlds, which Bull (2009) has described as signs of the hyper-post-Fordist moment of consumer culture, announcing the end of water-cooler and train conversations of which broadcasting material has always been a significant topic?

Reflection: Discussing TV Content

In your experience, to what extent are common TV programmes still discussed at the water cooler, in trains or in other such moments? In other words, to what extent do we still talk about commonly viewed media content?

Hence, one of the key challenges for radio and TV channels today is to continue to invite their users to participate in communities, share radio and TV moments, watch and listen live, here and now. Newer programme genres and formats that are built on reality events, such as *Big Brother*, that generate talk before, during and after the event (Scannell 2002), are certainly successful formulas that seem to have become the new mass moments of audience participation, both offline and online (see also Chapter 4 on transmedia storytelling and the 'Interactive branding strategies' box). Yet, the more traditional programme genres and formats that align to the daily rhythms of TV and radio consumption and co-construct the everydayness of these media (Moores 2005), such as news bulletins or soap operas, also remain important appointments for larger communities. Likewise, the mass ceremonies of watching simultaneously and en masse royal weddings, with the marriage of Prince William and Kate Middleton as an example, and sporting events, such as the Tour de France, are still far from dead and keep on contributing to this deep-rooted sense of connectivity, national, post-national or subnational identity (Dayan 2009: 21). Although the screens and devices through which people are taking part in these events are far removed from the old TV and radio sets, channels and stations continue to play an important role in this kind of reporting.

This highlights another risk linked to the for-anyone-as-someone structure. Rather than seeing the on-demand technologies as a liberation from the constraints and control of channel flows, they fear 'unprecedented degrees of selective avoidance' (Rosen 2004: 67) enabled through a thoroughly personalized and customized pursuit of one's personal taste. In particular, the much more rampant availability of entertainment material makes critical commentators conclude that interest in and responsiveness to informational, educational and cultural content will decline (Golding 2010: 214). Media users are less challenged and made 'incapable of ever being surprised' (Rosen 2004: 52), as the on-demand technologies do not 'encourage the cultivation of taste, but the numbing repetition of fetish' (op. cit.: 52). By tracking audience's tastes, these new technologies only surprise media users with programmes, series, music and news items they already enjoy, or at least consume. 'I don't want to have to listen to a song I don't like' (Bull 2009: 87–8), a quote of an iPod-user, captures the cultural Zeitgeist that in many respects seems incompatible with the blueprint of channels and stations, and with CDs or LPs.

Clearly, delivering and accessing radio and TV content dislocated from this channel-flow environment raises questions over how channels perform their role as 'judges', 'sorters' and 'providers' of content, with an eye to their audiences' tastes, wants and needs, and their own greater rationales (see, for example, Bennett 2008: 161). For example, public broadcasters bear a social responsibility to provide programmes that meet particular ideas about good quality, cultural identity, moral values and democratic citizenship (Murdock 2004). Historically, they have also played an important role in the creation of communities, feelings of national identity and bringing their audiences in contact with content that has educational, informational and cultural aspirations. Whereas the surveillance of public service broadcasters and other channels over their audiences has been the target of criticism, at least it was clear who was in charge of managing the flows. The new players that today are controlling what audiences can listen to and watch are far more invisible and unaccountable. Control has shifted to an independent sector in which computer programmers and internet entrepreneurs lead the way. Software tools and algorithms such as applied metadata protocols, filters, search engines and adaptive agent technologies, are also labelling, categorizing and valuing content, and hence monitoring and determining of what programmes and music we eventually engage with (Gripsrud 2010c; Uricchio 2004b: 252).

Chapter Summary

■ Because of the historical role that radio and TV channels have played as broadcasters, this chapter started with deconstructing the notion of 'broadcasting' and identifying its constituent features.

■ We then charted the historical stages that channels went through, drawing attention to the newest types of channels and the ways in which old channels are responding to the post-linear environment of which they are part.

■ In the next section we examined more closely the branding strategies that channels have been developing and applying to cope with their new competitors, and strengthen their image of trust and reliability in the landscape of content abundance.

▶

■ The second half of the chapter examined the traditional role of channels as gatekeepers and aggregators of quality content, and addressed whether they risk losing their meaning in the age of video streaming services like Netflix. Here, we critically observed audiences' engagements with television content from a global perspective, and concluded that a significant part of the discussion on channels' viability is biased towards the Western (predominantly US), industrialized, wealthy parts of the world.

■ Finally, we considered the democratic role of channels, and discussed a few authors who strongly argue for broadcasting channels as guardians of feelings of community, citizenship and the public sphere.

6 AUDIENCES IN THE DIGITAL BROADCASTING ERA

Compared to new media audiences, the audiences of broadcast media have been typically understood as 'passive' recipients of broadcast messages, with little or no room for active, interactive and creative engagement with the media content on offer. Since the mid-1980s, however, audience and television studies have consistently argued and demonstrated that the so-called passive audience was actively making sense of the media texts, creatively interpreting and understanding the media material and interacting with the offer by selecting and picking those programmes and channels that it really wanted to see and hear. The digitization of broadcasting, and particularly the convergence with the internet, has opened many more ways to intensify these germs of active, interactive and creative forms of media engagement. To some, we are witnessing the end of the 'old' audiences, defined as 'those who were on the receiving end of a media system that ran one way, in a broadcasting pattern, with high entry fees and a few firms competing to speak very loudly while the rest of the population listened in isolation from one another' (Rosen 2006) – they argue that for that reason, we should perhaps better speak about 'the people formerly known as the audience' (ibid.). Still, to others, audiences have not lost their relevance in contemporary media systems, because the various ways in which we can be an audience are essential to our social and cultural involvement in the world of which we are a part. How we understand ourselves and build our identities, how we are able to take advantage of the world around us, is intrinsically shaped by our audience affiliations and practices (Nightingale 2011a: 1).

Reflection: Identifying Audiences

For many people, the notion of 'audience' simply means 'people'. However, different media define their audiences distinctively. In what way do you see yourself as part of different audiences? In which context would you identify yourself as part of an audience of viewers, listeners, movie fans, gamers, downloaders, customers, citizens, device users or any other type of audience?

AUDIENCE APPROACHES

Audience research can be seen as a diverse and complex subject, and has resulted in numerous and often-conflicting theoretical approaches (Awan 2007). McQuail (1997) discerns three main schools of thought within audience analyses, which are also discernible in connection with digital broadcasting. First, there is the structural approach of audience measurement, most often starting from quantitative research. Besides measuring the size of audiences, this also involves charting the socio-demographic characteristics of audiences or users. This is typically linked to the industry perspective, where it is crucial for advertisers and channels to know how many people are listening to radio or watching television. From the industry point of view, audiences are conceived as a commodity that are bought by and sold to the advertisers (Ang 1996; Smythe 2006). Within the structural approach, audiences are often presented as monoliths, where viewing and listening rates are the ultimate criteria of audiences' experiences. With the advent of new delivery technologies (cable, satellite, VCR, internet), especially by the end of the 1980s (Ang 1996), the measurability of audiences has increasingly been put into question. In the age of digital broadcasting, measurements can be conducted on a much more detailed and finer-grained level, because user data can be collected via the set-top box and then be used for data-mining and audience profiling, as is already often the case on the internet (see also Chapter 3).

In the behavioural tradition of audience research, the second approach, attention is paid to the consequences of and reasons for particular media behaviours. Depending on the prevailing value judgements within research paradigms, the effect of the medium or message on the audience is seen to be either strong (passive audience) or weak (active audience). The first paradigm of the so-called 'effects' studies, also known as the hypodermic needle or magic bullet theory (referring to the metaphor of media messages being injected or shot into the audiences' minds), set the tone until the late 1960s. It mainly explored the question of how the media, television in particular, shaped the attitudes, feelings, thoughts and behaviours of the masses. From the late 1960s onwards, within US and UK mass communications studies, research interest shifted to the process of making sense and use of media content, emphasizing the social dynamics of viewing and challenging the pessimist view on media audiences' passivity (Nightingale and Ross 2003; Spigel 2004). Merton and Lazarsfeld (1968), for example, two illustrious media effects scholars, developed a 'content and response' analysis, in which they investigated the indirect effect of messages. Media effects were no longer considered as automatic, unavoidable and predictable, and individuals were assumed to be able to read media texts in a way totally different from the intended message of the producers. One notable behavioural research tradition that exemplified the shift towards the active audience paradigm was the uses and gratifications research tradition, which departs from the assumption that people make use of the media to gratify their needs (Blumler and Katz 1974).

The third, culturalist approach to media audiences also supported the active audience paradigm, but equally reacted against the individualist ideas about media use as advocated by the uses and gratifications tradition. Stemming from the Birmingham Centre for Contemporary Cultural Studies (CCCS) in the UK, and building on the work of Williams, Hoggart and Hall, cultural studies have argued that media audiences are culturally constructed and that they draw upon the socio-cultural contexts of which they are a part when making sense of media and popular culture (Scannell 2007). That is to say that audiences do not take media for granted, but interpret them starting from their everyday social situation and status (Nightingale and Ross 2003; Scannell 2007). In the early 1980s, this tradition gained momentum with qualitative audience research (e.g. interpretative research, ethnographic research, reception research) driven by, in Morley's words, the question of 'how particular people, in particular contexts, perceive the relevance (or irrelevance) of specific media technologies for their lives, and how they then choose to use those technologies or ignore them, or indeed "bend" them in some way, to a purpose for which they were not intended' (Jin 2011: 128).

UNDERSTANDING NEW TV AUDIENCES: TOOLS, AFFORDANCES AND PRACTICES

Whereas broadcasting audience studies – which have boiled down pre-eminently to television audience studies (the field of radio audience studies is remarkably smaller) – have been typically oriented to studying the response to and engagement with traditional, centralized, one-to-many broadcasting systems, the transition towards digital broadcasting, 'me-casting' and 'peer-casting' (Merrin 2008) and ubiquitous media clearly require the integration of insights coming from the research field on ICT, digital media and digital culture. Dovey (2008: 244) argues that the technological developments of the twenty-first century have created areas of tension between 'old' and 'new' ideas about people's engagement with the media. Rather than speaking about *active audiences*, we are dealing with *interactive users*; *interpretation* is making room for *experience*; *spectatorship* is challenged by *immersion*; *consumers* are becoming *participants* and *co-creators*; what used to be *work* is today *play*. Drawing upon the fields of ICT research and Science and Technology Studies (STS), the interrelated notions of 'tools', 'affordances' and 'practices' are here helpful to understand how and why audiences are appropriating digital broadcasting in their everyday lives (Pierson 2003: 606–7; Van den Broeck et al. 2006).

First, the notion of tools refers to the technological media that are being used to access broadcasting content. Radio and television have both evolved into technological artefacts that are omnipresent and completely taken for granted in people's homes worldwide. They have become domestic and domesticated technologies (see also Chapter 2) (Hartley 1999; Silverstone 1994b; Silverstone and Hirsch 1992). We do not see them as technologies anymore, but as something we have tamed and has become natural and everyday. In particular television has developed into a pivotal domestic medium, which is shown not only in the time spent on it – watching television still dominates our free time at home – but also in the place it is given in the home – often centrally positioned in the living room, surrounded with ornaments, or as the pièce de resistance itself (Haddon 1992; McCarthy 2000; Silverstone 1994b). One explanation for television's centrality in people's everyday lives is that from the mid-1950s onwards, using television required little technological competence, to the extent that the artefact became, in Heidegger's terms, 'ready-to-hand' (Dourish 2001).

Today not only is the television set omnipresent, but related (digital) technologies, also denoted as 'TV-extensions', have equally found their way into many households in the industrialized world (Intel 2005; IP Network 2008). Digital broadcasting tools are materialized in various applications, devices, interfaces, platforms and services: movies and entertainment programmes can be viewed in a HD-screen home theatre setting with surround-sound systems and ambient lighting; news flashes stream in on mobile phones; people can opt for time-shifted viewing with the personal video recorder; viral ads enter via YouTube on computer and tablet screens. People's former experiences with and their capability to handle these tools are crucial for understanding their engagement with digital broadcasting. For example, people who are more computer literate will probably more easily watch television programmes via the internet.

Reflection: A Multitude of Broadcasting Tools

The tools or technological devices for consuming broadcasting have increased substantially in the era of digital broadcasting. Consider your own situation and the devices you (or your friends or family) use to access, record or share television or radio programmes. What kind of changes have you experienced over the years?

Second, the notion of affordances, originally coined by the perceptual psychologist Gibson (1977) and later introduced in the field of computer science and human-computer interaction by Norman (1988), expresses the idea that each artefact consists of properties, i.e. design aspects, that suggest or determine how that artefact could possibly be used. So, '[A] chair affords ("is for") support and, therefore, affords sitting. A chair can also be carried.' (Norman 1988: 9). In order to understand the use of any technology or medium, the possibilities it offers, and the meanings and expectations attached to it, need to be understood. Technically an affordance is seen as a property of the environment that affords action to appropriately equipped organisms (Dourish 2001: 118). Taking the example of the chair once again, it is clear that the chair affords sitting to human beings, because its seat fits the human body, but it does not afford sitting to an elephant, as this organism is not 'appropriately equipped'. In this way, an affordance is a three-way relationship between the environment (or system), the organism and an activity: the environment is perceived in terms of its potential for action by an organism.

Third, the notion of practices, borrowed from the fields of anthropology and ethnography, suggests that material artefacts and goods are important, yet not for their own sake but for the practices they make possible. Also within the field of design research, it is argued that we should 'think of products in terms of verbs, not nouns: not cell-phones but cell-phoning' (Kelley and Littman, 2001: 47). The relation between product and practice is dynamic, meaning that they co-evolve. This is in line with the social constructivist perspective stating that there is no essential use to be derived from the artefact itself, because technologies should be studied in their context of user practices, and users and technologies should be seen as co-constructed (Oudshoorn and Pinch 2003). Practices exist as 'recognizable entities' (Hand et al. 2005), as 'forms of bodily activity, forms of mental activity, things and their use, background knowledge in the form of understanding, know-how, states of emotion and motivational knowledge' (Reckwitz 2002: 249). As they 'require constant and active reproduction or performance' (Hand et al. 2005), practices are seen as routinized types of behaviour. Within social theory, and communications theory in particular, 'practice theory' studies media uses in terms of everyday, habitual and routine media practices, of doing things with the media, rather than interpreting media texts (Couldry 2004b, 2011; Moores 2012; Reckwitz 2002). Depending on the background, daily life context, the competences and communities users can draw upon, practices will differ. Hence, there is no such thing as 'the' user; rather we are dealing with different users and user communities as well as with the same user in different settings (at home in the living room, in the office at work, on the train, in the pub, etc.). In addition, user practices will vary according to the kinds of tools used to access the different types of broadcasting content, often platform-independent (see Chapter 3).

Although the technological changes broadcasting is going through seem to bring about various new kinds of affordances, many of these new affordances were already envisioned in 1883 by the French futurist novelist Albert Robida. He wrote about the 'telephonoscope' as a site of news, home entertainment, surveillance, person-to-person communication and public information (Uricchio 2009: 71). Starting from the domestication of TV in the home (see also Chapter 2), we can connect (digital) broadcasting's affordances with the four functional domains typically allocated to the household (Zerdick et al. 2000: 214–17), that is: (1) entertainment and relaxation, (2) news and commentaries, (3) the organization of daily life, and (4) interaction with social networks or sociality. Each of these four domains has been part of broadcasting's logic, as central affordances for audiences. Traditionally, analogue, linear, channel-organized television was placed at the crossroads between 'news and commentaries' and 'entertainment and relaxation', and intrinsically, in its characteristic of providing routines and a time structure through the programme schedule, it has also been crucial for the organization of daily social life. The key question today is whether and how media audiences will accept, resist or reinterpret the changes their trusty television screens and sets are going through, and how this might show in their practices. In what follows, we will discuss for each domain the changing affordances of broadcasting technologies and how these will or might interact with other forms of audience practices.

AFFORDANCES AND PRACTICES OF ENTERTAINMENT

Television is traditionally regarded as 'a medium for passing time, for being amused and for taking it easy' (Pool and Noam 1990: 241). The huge popularity of television can be explained by appreciating that it provides distraction, recreation and relaxation at a low cost, and with little effort on the part of viewers – which makes it an ideal tool for 'mood improvement' (Lee and Lee 1995: 13), that sooths rather than stimulates (Mullan 1997). Watching TV is a culturally accepted rest activity (Helle-Valle and Stø 2003), well-delineated in time through the 'appointments' with programmes at fixed hours. In its affordance of offering relaxation and 'social inactivity' (op cit. 48), television is often described as a 'lean-back experience' – in contrary to the 'lean-forward experience' of the computer – and a typical characterization of the relaxed viewer is that of the 'couch potato' (Buonanno 2008; Lee and Lee 1995). Although viewing figures seemed to support the argument that 'what audiences want mainly from their television is not information but entertainment' (Lull 1990: 155), the viewing audiences did not always derive pleasure from the entertainment they were offered (Bauwens 2002; Pauwels and Bauwens 2007). Yet with the constant and abundant availablility of on-demand content in the digital era of broadcasting, and the possibility to use television for other computer-related services as well, the opportunities for having a gratifying TV experience have increased. Just as the iPod has put an end to the tyranny of radio playlists, where you had to listen to songs you didn't like (Bull 2009: 87–8) (see also Chapter 5), the digitization of television opens up a large realm of entertainment from which you can select and choose. Hence, the linear activity of just sitting and watching TV is moving towards an interactive engagement with the television programmes on offer by making deliberate choices about which shows, series and films to watch, and when. And even the engagement with the programmes themselves, as the popular reality TV formats of today demonstrate, is built on 'the industry's current ideal of an interactive, self-synergizing media consumer' (Hassoun 2014: 272) where trans-media audiences are urged to vote, tweet, like and share via the other platforms (see Chapter 4).

These new forms of engagement with television are shown in the way 'old' practices of 'focused viewing' and 'background viewing' (Hamill 2003) have been redefined. In the early days of television, when broadcast hours were still limited, people absorbed the television spectacle by settling down in front of the TV set and watching the show in a focused way. This mode of watching TV, also described as 'TV in the front'-use – in which watching television is the primary activity (Van den Broeck et al. 2006) – has not completely disappeared, but it has been argued that the focused mode of watching is not compatible with the medium of TV, which, unlike cinema, does not demand extraordinary effort or concentration (Ellis 1982; McLuhan [1964] 1994). Hence, with the proliferation of channels, programmes and programming hours, and with the permanence of and audience's familiarity with TV, the audience's engagement with TV has moved gradually towards a more background use – like radio, as a matter of fact. Hence 'TV on the side'-use (Van den Broeck, Lievens, and Pierson 2006) – in which watching television is given second place, while conducting other activities at the same time – has increased. TV offers a kind of absent-minded distraction, without compromising the main task that is simultaneously carried out. Audiences' familiarity with and understanding of the medium's predictability in terms of conventions, structure, pace, cues, modes of address, etc., have made TV a popular side activity next to other pursuits. So the practice of dividing ones attention between television and other parallel activities has become a common practice (Buonanno 2008). Interestingly, industry aims at capitalizing on this mode of watching 'TV on the side', by combining audience's attention with the multiplatform engagement many programs of today have built in (see also Chapter 4). Hence, producers and broadcasters are encouraging people to use other media, screens and communication technologies 'on the side' while watching the show, but at the same time trying to keep their audiences' mind on the show.

'TV in the back'-use (Van den Broeck, Lievens, and Pierson 2006), also described as 'ambient watching' (Barkhuus and Brown 2009: 14), in which television becomes a 'environmental resource', 'companion' (Lull 1990) and 'electric light' (Lee and Lee 1995: 11), with the medium functioning as wallpaper and background noise, is another important viewing practice, in particularly among young adults (Peters 2003). As often argued, these different modes of paying attention to television do not exclude one another, but are part of many people's everyday use of TV, depending on the programme they are watching, the company they are in, the cultural context they are part of, or the time of day (Fiske 1987: 73–4; Lee and Lee 1995; Lull 1990).

Reflection: Engaging with Television and Entertainment

Consider your own experience when watching television for entertainment. Can you identify different levels of engagement in your own viewing behaviour, and on what these are dependent (e.g. content, context, other people present)? Do digital broadcasting services like pausing live television and recording programmes have an influence on the way you watch television?

AFFORDANCES AND PRACTICES OF NEWS

From its inception until today, broadcasting media have played an immensely important role in the dissemination of news, the democratization of the public sphere and the social involvement of citizens. Public as well as commercial broadcasters have always offered a large number of news programmes, talk shows, current affairs magazines, etc. that rank high in viewing figures and appraisals (Bauwens 2002; Mullan 1997). In particular, a substantial democratic potential has been attributed to radio and TV news, as it has always been seen as the pre-eminent form of broadcasting output that explicitly addresses the audience as citizens. As such, the practice of tuning in to the news is perceived as taking a step outside the limited circle of one's everyday life, which makes the person (feel) part of a larger whole, and in this way fulfils an act of political participation or citizenship (Bauwens 2002; Vandenbrande 2002). Even though people do not always watch the news that attentively, and are much more loyal to entertainment programmes (like soaps, quiz shows, etc.), many still share the 'modern' belief of journalism's prominence for democracy and for news broadcasting in particular (Costera-Meijer 2004). Television news, specifically, occupies a prominent place in audiences' media news hierarchy, as it is regarded as the ultimate 'essential' news review, and therefore has the strongest 'truth' status for audiences. Audiences value the television news for enabling involvement with (distant) events that happen outside their personal sphere (Bauwens 2002: 443–7). If it used to be a fixed routine – watching the evening news has for a long time been an established tradition in many households, indicating also the start of prime-time television – the practice of consulting news has to compete with many other daily activities in private life, and with many other easy-to-use and on-demand news media. Even with the current proliferation of digital devices, television news remains the main way of getting the news in many countries, whereas network TV bulletins and 24-hour TV news channels remain the most popular platform for breaking news (Newman and Levy 2014: 44).

However, internet and mobile technologies have come to play an increasingly prominent role, especially with regard to the speed and interactive character of news (cf. Schrøder and Larsen 2010). Hence, websites and apps are taking second place, before radio and social media. Obviously, audience socio-demographics, contexts of use and type of news content are crucial determinants of how people are making use of traditional (TV, radio and printed newspapers) and online media (websites, apps, search engines and social media) to access news. For example, young people aged 18–24 are twice as likely to use online in comparison to traditional sources, while for people aged 55+ it is the reverse. In the personal spaces of the home (like the bedroom and study), people aged under 45 are almost three times more likely to access internet-based news than watch TV. Even in the communal spaces of the home (like the living room and kitchen), internet-based access to news now more or less equals TV-based access. Although tablets and smartphones are more common, both the smart and the connected television are new, upcoming internet-based devices which are also used to access news (Newman and Levy 2014: 45–8, 64, 83–4).

In addition to the coming of internet and digital media, the democratic potential of news has expanded. People are not only passive consumers of news anymore; they are also increasingly able to produce news themselves (e.g. bloggers, citizen journalists). Hence, when looking at the research on changing news media, we see a shift from an imaginary news public to an audience that increasingly takes an active role, and from traditional media to new media (Picone 2010). The latter is demonstrated by the success of social sharing sites for news (like Huffington Post, Buzzfeed, Upworthy, etc.), news aggregators (like Google News, Yahoo, MSN, etc.) and other online news initiatives.

Reflection: Consulting News

The practice of consulting news (as with any other broadcasting content) has become enormously diversified. Think about how you yourself consult news. How do you value television news in comparison to other news genres? In what way have the internet, social media and other digital channels changed the way you consume news?

AFFORDANCES AND PRACTICES OF EVERYDAY LIFE ORGANIZATION

The broadcasting flow system has developed into specific listening and viewing patterns, rooted in people's daily lives (see Chapter 5). As Hamill (2003) notes, when television was introduced it had a special place in the daily routines and patterns, but 'it appears to have become absorbed into other patterns, reflecting what we call the rhythm of daily life' (Hamill 2003: 72). These fixed use patterns can be partly ascribed to the so-called 'appointment model' of traditional broadcasting media, which allowed audiences to tune in to channels, but not to influence programming hours. The phenomenon of serialization, such as the daily or weekly recurrent episodes of shows, exemplifies this model (Barkhuus and Brown 2009; Buonanno 2008). As a result, viewers became used to the fact that when they wanted to watch a programme, they had to be ready when it started in order not to miss it. Hence, the organization of everyday life became a central affordance of television, which gave rhythm and structure to people's lives in terms of when and what to watch. Even with the introduction of the VCR and pay-TV, viewers were still dependent on the programme offer and start hours as set by the broadcasters.

Today, people have access to a multitude of platforms and make increasing use of additional devices that enable them to time-shift or watch content other than the live broadcast stream. Indeed, research demonstrates that time-shifted TV usage is increasing, especially among younger generations (Rideout et al. 2010) – which might eventually lead to the dethronement of TV as 'the tribal drum' (cf. McLuhan [1964] 1994) that supplies the beat for everyday social life. However, the phenomenon of a broadcasting menu that is fixed by the broadcasting companies is still prevalent, and some studies indicate that old habits of viewers and listeners die hard (Simons 2013; Van den Broeck 2010). Taylor and Harper (2003) found that in the time-span when (working) people watch television, with the prime-time hours from 6–7 p.m. to 10–11 p.m., 'regular patterns of viewing' transpire. The coming-home viewing, for example, is aimed at relaxation: 'switching on' the television 'to switch off' the call of duty. This practice of viewing is highly disengaged and the selection of content to watch is rather unplanned, with a lot of zapping or channel surfing taking place. In the mid-evening viewing, the typically planned viewing behaviour comes to the fore, with higher levels of engagement and the so-called practice of 'viewing by appointment' of favourite content such as news, soaps and game shows. This viewing is mostly social viewing and other activities (e.g. dinner, homework) are geared to and marked by it (Gauntlett and Hill 1999). In the later-evening viewing (starting about 9 p.m.) household chores are finished and people tend to have specific programme preferences (e.g. documentaries, dramas and current affairs programmes), which they watch with a high level of engagement. In this stage, programme guides are being used more often and the active selection of programmes takes place (Taylor and Harper 2003).

One of the key paradoxes about watching television is that it is not considered a priority activity, but at the same time, it is one of the most pursued leisure-time activities. Although marketing studies show that traditional TV viewing is systematically declining among young people aged 18–24, they also point out that most adults still watch a lot of broadcast TV, and most of the time they watch it live or scheduled (see for example www.marketingcharts.com; Consumerlab – Ericsson 2013), as part of their daily routines. For example, in Flanders (Belgium), more than 75.8 per cent watch more than one hour per day during the week, and 81 per cent during the weekend (De Moor et al. 2013). Overall, most time is still spent on media at home, with linear television as the predominant video platform (Taneja et al. 2012: 964). Although empirical evidence shows that TV is still present in the media life of tweens (9–12 years old) and teenagers (13–16 years old), even in countries with a high penetration of online and mobile technologies – such as Sweden (Westlund and Bjur 2014) – young people's engagement with TV is foretelling the movement from appointment television to 'networked video culture' (Marshall 2009: 41), where different television extensions and mobile devices are challenging the concept of punctual, routinized television viewing we have known since the 1950s (Horowitz Associates 2014). Does this also mean that the structures and uses of broadcasting, which have always been so crucial to the practical organization of everyday life, have become irrelevant? It is still too early to know for sure, but research on cross-platform media use shows that although digital media enable 'anytime, anywhere' consumption of content, the day-to-day rhythms of people's lives, which are shaped by work, sleep, family, leisure, commuting and housekeeping, still play a decisive role in how and which media are used. Audiences keep relying on 'habit and iteration in creating their media repertoires' (Taneja et al. 2012: 964). Future habits of media use and of TV consumption in particular will demonstrate the ways in which new modes of television viewing will reproduce and reconfigure people's media practices in relation to the organization of everyday life. Some believe that structural constraints, like patterns of class consumption and distinction, weather conditions and housing cultures will keep on determining our engagement with TV – in both its symbolic (what significance the medium has in constructing our and our families' identities) as well as its tangible sense (as to where to put the TV set) (Olofsson 2014).

AFFORDANCES AND PRACTICES OF SOCIALITY

Broadcasting media have been regarded as tools that afford sociality for everyday life in a modern, large, anonymous and fragmented society. This can be traced back to the installation years of broadcasting, when people from different households met around the first available radio and television sets in the street, to listen and watch together (see also Chapter 2). In the early days of television, when it was domesticated, the home became a unit of solidarity, as family members gathered around the TV set, the 'electronic hearth' (Tichi 1992), to watch programmes that were intended for family viewing (Flichy 2007; Livingstone 2009). So watching television was a family event that occurred in a domestic and informal context, often accompanied by conversations about the programmes, themes and people shown on TV, both inside the home with family members and outside with friends and colleagues (Burton 2000; Katz et al. 1974; Morley 1986; Seiter 2001).

Case Study 6.1: Social Uses of Television

Lull (1990) identified a typology of the social uses of television, which can be seen as the different ways in which the affordance of sociality of broadcasting is expressed. He distinguishes structural and relational uses of television. Structural uses comprise the use of television as 'an environmental resource', which provides background sounds and images, companionship during other activities and entertainment. The use of television as a 'behavioural regulator' is a second type of structural use. Television regulates social routines (e.g. dinner time, bedtime and other activities or duties). This is connected with the earlier affordance of the organization of everyday life.

Relational uses of TV refer to 'the ways in which audience members use television to create practical social arrangements' (Lull 1990: 37). Relational uses comprise four dimensions or functions: communication facilitation, affiliation/avoidance, social learning and competence/dominance. First, the content of television programmes offer viewers a common ground for talking, or, for example, can illustrate an experience or lead to a discussion about values when a controversial programme is aired. Second, the affiliation/avoidance function is related to the physical experience of watching TV, where people use TV, for example, to sit close to their relatives and induce conversation. The most important function in this is the creation of a shared viewing experience. It can lead to discussion, but also provide a 'natural' and socially acceptable way of being silent, all focused on the same programme. In this sense television can also serve to reduce conflict. Third, viewers can use television as a means of social learning about how to behave in social life situations. Fourth, television can be used for role enactment or reinforcement, the exercise of authority or gatekeeping in order to manifest competence and dominance.

Some authors proclaim that the personalization and privatization of television use are eroding the social affordance of television. As early as 1972, Bogart had already noticed a trend towards more personal viewing patterns because of smaller families and lower receiver costs, both leading to a smaller number of viewers per receiver. This trend was reinforced in the 1980s and 1990s, because of among other things the rise in single-person households (Fowles 1992). More contemporary scholars like Katz (2009) argue that the television of 'sharedness' no longer exists. Given the rise in available channels and the changing television system (evolving towards the internet and other media), the old idea of TV as nation-builder and promoter of family togetherness belongs to the past (see also Chapters 2 and 5). The status of television as centralized around 'events' that were shared amongst the nation has lost its validity. Within families, just as the portable radio opened the door to personalized uses of the medium, television is, in Putnam's (2000) terms, losing its bonding role of strengthening close-knit relationships. In particular, in families with teens, the commonly shared television experiences are more rapidly crumbling than in other families (Westlund and Bjur 2014: 25).

Reflection: Affordances of Digital Broadcasting

Radio and television broadcasting have traditionally provided entertainment, news, everyday life routines and sociality for people. In the digital broadcasting era these properties could be reshuffled and/or extended.

a) Are these the kind of properties you also experience in your private life?
b) Do you find other characteristics being (more) important for you?
c) Did you experience a change when switching from analogue to digital television?

Audience studies, from the historic to the recent, demonstrate that people do generally prefer to watch television together (Barkhuus and Brown 2009; Lee and Lee 1995: 12; van Zoonen and Aalberts 2004: 12–13), but that the function of television as a platform for conversations and discussions within a family is thoroughly minimized by people themselves (Bauwens 2002). Drawing on Lull's (1990) study of family viewing, it is clear that the social affordances and practices of the medium within the household interact with the 'complex mixture of people, social roles, power relations, routine activities, processes of interpersonal communication, ecological factors that characterize the home environment, and technological devices and appliances that exist there' (op. cit.: 159). Hence, in some families, in certain stages of couple and family life, television can be used to avoid conversation (Bauwens 2002: 485; Gauntlett and Hill 1999). In other families, the TV set sometimes even causes tension within the family, and leads to barely acceptable compromises about what is watched (Flichy 2007; Gauntlett and Hill 1999). Still other people point to the forced sociability that TV brings about. Precisely because many families regard television viewing as a family event, members who decline to watch TV and therefore decide to leave the living room to do something else are seen as spoilsports or perhaps even deviant to family norms. For this reason, these people frequently watch programmes they actually do not wish to see, but are willing to shelve their own preferences if it entails watching along with the others (Bauwens 2002: 282, 479–80) – or, to rephrase it in Lull's (1990: 148) words, the practice of 'television viewing is constructed by family members, it doesn't just happen'.

If the social nature of traditional television experiences can be questioned, at the same time one can also wonder whether there will be a linear evolution in the direction of individualized 'me-TV' in a digital broadcasting environment. On the one hand, this seems to be the case, given the more convenient ways to control and individualize television watching (e.g. via DVR), or the rise in the number of personal appliances to access TV content (e.g. smartphones, tablet computers, etc.). On the other hand, digital broadcasting can also offer new ways to socialize by being in contact and interacting with the programme and/or with others who are watching the same content elsewhere (see also Chapter 4). For example, in Flanders (Belgium), 9 per cent of all TV viewers use the internet to give their opinion or talk about a programme. However, this figure is much higher in the age group under 30 (32.2 per cent). This kind of interaction mainly takes place via mobile devices (smartphones, laptops and tablets), with the most common channel (91.4 per cent) being social media such as Twitter and Facebook (De Moor et al. 2013). Hence, the 'water-cooler talk' people have the day after a programme is aired is now complemented by 'instantaneous water-cooler-talk away and apart from the television screen' (Johnson 2009: 115) – and is increasingly encouraged by computer and mobile content streams. Based on the combined play of 'despatialised simultaneity' (cf. Thompson, 1995: 32) – i.e. the original affordance of watching the same programme at the same time from different locations, which has always been one of the fundamentals of broadcasting – and 'remote connectivity' – i.e. the old affordance of telecommunication enabling technologically mediated person(s)-to-person(s) communication – these new forms of television audience formation are also known under the umbrella term of 'social television' (Ducheneaut et al. 2008). Through 'systems that allow remote viewers to interact with each other via their television set' (Geerts 2009: 2), audiences' recommendations, commentaries, links, feedback, file sharing, friendship groups, micro-blogging, status updates and the various social media, not only provide the connective tissue, but also the content for television (Bruns 2008; Goggin 2011; Schatz et al. 2007) (see also Chapter 4). Some argue that this kind of sociality exceeds the scope of the fan cultures previously supported by television magazines and websites; and goes beyond the wider user selection of television viewing now enabled by downloading using peer-to-peer programmes (e.g. BitTorrent) (Goggin 2011).

Chapter Summary

■ In this chapter we discussed how audiences are being redefined in the digital broadcasting era. First, we examined how audience research consists of numerous and often-conflicting theoretical approaches.

We elaborated on three main schools of thought within audience analyses, which are also discernible in connection with digital broadcasting: (1) the structural approach of audience measurement, (2) the behavioural tradition and (3) the culturalist approach.

■ In order to understand potentially new perspectives on television audiences in relation to technological changes, we took a socio-technological viewpoint. In particular the interrelated notions of 'tools', 'affordances' and 'practices' are helpful to comprehend how and why audiences are appropriating digital broadcasting in their everyday lives.

■ Starting from the domestic settling of TV in the home and its surrounding practices, we then connected (digital) broadcasting's affordances with four functional domains typically allocated to the household: (1) entertainment and relaxation, (2) news and commentaries, (3) the organization of daily life and (4) interaction with social networks or sociality.

■ The key questions were whether and how audiences (will) accept, resist or reinterpret the changes their trusty television sets and screens are going through, and how this might interact with their practices.

7 RETHINKING DIGITAL BROADCASTING AND NEW MEDIA

Drawing upon Dahlgren's (1995) metaphorical imagery on television, we have taken a multifaceted approach in this book and have treated digital broadcasting as, in Dahlgren's (op. cit.) words, a prism with many sides, which one has to consider from different angles in order to catch the meaning of digital radio and TV. Hence, digital broadcasting is 'simultaneously an industry, sets of audio-visual texts, and a socio-cultural experience' (Dahlgren 1995: 25). In Chapters 2, 3, 4, 5 and 6 we have tried to explore these three facets from, respectively, a historical, political-economical, production, channel and audience perspective. Throughout the chapters it has become clear that even if we focus on one side of the prism, the other angles also need to be taken into consideration in order to grasp more fully the meaning of today's and future broadcasting. That is why audience practices cannot be dealt with without referring to political-economical configurations; aspects of production cannot be understood without technological developments and socio-cultural transformations; the role of channels is closely related to policy issues; etc. It can be argued that a prism inevitably causes 'difficulties of seeing all the sides at the same time' (Dahlgren 1995: 25). However, in this last chapter we conclude our quest by drawing connections between an amalgam of interrelated processes that express the dynamics that digitization has accelerated, if not started. The proposed framework, which is definitely not exhaustive and determinate, highlights some key ideas and critical observations transpiring from the current field of research that is following and scrutinizing the changing ontology of broadcasting media today and in the near future.

LIQUIDIZATION

The sociologist Bauman (2000) has argued that fluidity or liquidity is the leading metaphor for modern society in the twenty-first century. The characteristic features of today's social institutions, phenomena, practices, communities and identities are their mobility, inconstancy and instantaneity. In such a world, the 'flow of time' and 'aggregate of moments', in Bauman's words, count more than the solid spaces and places where the social, in all its meanings (i.e. life, practices, organizations, etc.) is settled. In line with this analysis, one could state that broadcasting, as we know it today, is a critically liquid phenomenon and a truly fluid technology and institution. In the late 1990s sceptics still argued that the speed of change and breakthrough of digital TV was overestimated and that 'plain old television as a proven medium' (Steemers 1997) would never lose its importance. Yet today it is widely acknowledged that, for example, the worldwide appeal of YouTube and the prevalent culture of video production and consumption, or the impact of internet, on-demand and participatory media is heavily affecting the old modus operandi of broadcasting media. In Chapter 2 we saw examples of how broadcasting media and television in particular have always been in a state of flux and have continuously changed their shape in order to adapt to market conditions, which has probably improved their chances of surviving (Küng et al. 2008: 159). Media historians, too, have extensively demonstrated that radio and television were never static media, but have gone through several transitions, which were also considered as being disruptive in their times (Uricchio 2013). However, in the age of multiple media platform production and consumption, technological convergence and economic mergers, the chapters of this book have demonstrated how the well-delineated definition and categorization of radio and television broadcasting is arguably more unsettled than many researchers could have foretold.

For a start, broadcasting media are no longer entities in their own right with separate media features, economic logics and social practices. Instead, they are increasingly intertwined – converged – with digital media, whose boundaries are constantly being redrawn (Taneja et al. 2012; de Valck and Teurlings 2013; Westlund and Bjur 2014). Our everyday life consists more and more of densely available, increasingly mobile, computational and communicative resources that are assembled in heterogeneous ways (Dourish and Bell 2011). To put this in perspective, people have been using multiple media in their everyday lives for quite some time. Even the phenomenon of media-multitasking, or more specifically simultaneous media use, is relatively old, since older generations also combined multiple media in their everyday life, as they used to read the newspaper while listening to the radio or watching TV (cf. Voorveld and van der Goot 2013). However, each of these old media contained specific content, which one could only find, enjoy and engage with within the contours of these media. What is more, each medium accorded with or was confined to a particular time and place: we read the newspaper in the morning at breakfast, on the train; listened to the radio in our car, at work; watched television at home before or after work. Each medium also made an appeal to one or more of our sensory faculties: print media to our sight, radio to our hearing, television to our sight and hearing. In the digital era all these boundaries have become blurred. The conventional uses of the media (TV is for watching, newspapers for reading, radio for listening), the interconnection between media and particular content (soaps are on TV, hit charts on the radio, opinion pieces in the newspaper) and the long-established lodging of media in particular times and places are all evaporating. Media historians and media theorists have pointed out that this process of liquidizing is the inevitable acceleration of much older developments which go back to the early modernist projections of media and society. However, when it comes to media uses we are probably seeing only the beginning of what one might call 'liquidizing audience practices'.

For example, one of the key questions within both academic and industry-driven audience research is how people are moving and switching from one medium platform to another in their daily life rhythms and how they make use of multiple media platforms, often handheld devices, to engage with the particular content of their preference (Taneja et al. 2012). From a methodological point of view, one of the major challenges remaining, especially for academic research that is still heavily dependent on industry-driven data on this matter (like, for instance, Nielsen) (cf. Hassoun 2014), is how to observe and register cross-platform media use and new forms of television viewing (Wood 2007). It is argued that the accuracy and reliability of self-reports, often used in audience studies, is considerably undermined by people's poor ability to recall which media they have used, given the abundance of delivery technologies they now have at their disposal (Taneja and Mamoria 2012). If we want to learn more about the importance of broadcasting media in audiences' larger media repertoires, and take a more thoughtful stance in the debate on change and continuity, it will be crucial to develop more fine-grained and longitudinal observation techniques.

DEFAMILIARIZATION

One of the main outcomes of liquidization is the 'defamiliarization' of broadcasting media. By this we mean that the digitization of television places this ordinary and mundane medium in a context that is unfamiliar and uncommon for people who saw TV entering into their homes (and/or have known many years) without the internet. Yet, unlike in art where the technique of defamiliarization invites the audience to become critically and reflexively aware of the familiar and taken-for-granted, audience studies show (and industries know) that the defamiliarization of TV is unlikely to totally revolutionize the socio-cultural experience of watching television. The degree of inertia in our viewing was discussed in Chapter 6 and is especially true for the older, pre-digital generations who keep on thinking about the medium of television broadcasting within the earlier paradigm of mass communication (Hermes 2013).

However, as shown at some length in the previous chapters, the changes in everyday televisual culture are subtle and incremental. And yet the unplugging of the television box from the act of watching TV – you no longer need television to watch TV, and you can do other things with your TV box, like playing Wii Sports – potentially preludes a rearranged televisual culture. This may well be more tactile (see Giddings and Kennedy 2010) – as already foreseen by media theorist McLuhan ([1964] 1994) – simultaneously dispersed over multiple screens and, last but not least, hybrid, as it merges offline, concrete and hence spatially structured viewing experiences with online, virtual hanging out in the vast space of audiovisual content. The discussion in Chapter 6 suggests that for many people these experiences are still not the norm when watching TV. However, these processes are already ceasing to be visible and novel among the younger generations, at least in the more affluent parts of the world (Jones and Fox 2009; Rideout et al. 2010).

STORYTELLING

As soon as the fascination with new, magical, shiny media and communication technologies has calmed down, we are ultimately left with the stories told and circulated by these technologies. From an industry perspective, it has always been clear that there can be no media without symbolic content, and, what is more, that this symbolic content is the foundation for making profit. In that respect, the 'Content is King'-quote, originally coined by Bill Gates in 1996 when the gold rush dot.com years took off and IT enthusiasts were exploring all the technological pathways that the internet had opened up, was perhaps visionary for all that has taken place since then (Gates 1996).

From a media studies' perspective too, it has been argued that good stories, gripping narratives, culturally relevant messages, engaging words, captivating games and appealing images are primordial to media. As Silverstone (1999: 41) sharply noted, our culture still preserves a profound sense of enchantment. And that is what media do, they enchant. They fuel our fascination with our origins and futures, as they are 'social text: drafts, sketches, fragments, frameworks; visible and audible evidence of our essentially reflexive culture, turning the events and ideas of both experience and imagination into daily tales, on big screens and on small' (op. cit.: 41). Therefore the story survives.

In Chapter 4 we saw how today's media culture, non-professionals and ordinary people are all contributing to what one might call the 'storisphere' (Mu et al. 2013). Yet, the need for stories that meet or come close to professional standards (for example, the work of pro-ams, amateurs who work to professional standards (Leadbeater and Miller 2004)) is not fading away, and one can look back on a long history of media professionalization (see Hermes 2013), in particular in audiovisual media production. Maybe we are watching, with all its technical limitations, news snippets on our laptop, the summary of a football match on a small handheld device, or a TV series video-streamed from the internet – which is steadily gaining in popularity with the increasing network bandwidth and the success of video-streaming websites like YouTube, Vimeo and Hulu (eMarketer 2010). But we are nevertheless watching professionally made television content, often produced within the institutional contexts of large broadcasting companies. Obviously, user-generated content and the culture of video interaction are affecting broadcasting journalism (see, for example, De Dobbelaer et al. 2013), television screenplays and camera work (see Chapter 4). Yet they are not ruling out the traditional role that broadcasters have always played in producing good entertainment and trust-worthy information, even for what one might call the 'second-screen, social-media generation'.

DE- AND RECENTRALIZATION

As argued by others and shown throughout this book, the breaking up of the sequenced, linear flow model of broadcasting is putting television's role as being central to our media cultures, lives and societies to the test. In particular we have claims about the so-called second-screen, social-media generation, habituated to and socialized into using 'à la carte' online media menus and fully familiar with the technologies of 'miniaturized mobilities' (Elliott and Urry 2010: 28). For this generation, will television step down as the primary screen? And yet researchers keep on pointing out that the younger generation will not stay young forever and their own media use will itself be susceptible to changing demands and interests over the course of their life (Westlund and Bjur 2014). The structural constraints of adult, everyday life, and its related responsibilities, duties and wishes (like a job, a sustainable romantic relationship, a harmonious family, etc.) as well as its everyday routines, invariably make up the framework in which people will decide which media platform, screen or technology to let into their lives. Moreover, it is precisely these non-media aspects of our lives (where and how we emotionally and materially organize all aspects of our life) that do not change that quickly, in contrast to the way in which the technologies we are confronted with evolve at an increasingly rapid pace.

Despite the potential new forms in which digital broadcasting could develop, both broadcasters and television production industries are also actually, and intensively, searching for ways to reinvent the mass communication paradigm of analogue broadcasting. Even though channels are making every effort to become part of the larger multipoint networks of video traffic, they are at the same time developing strategies to differentiate their own audiovisual content, which they produced or sourced, from the mass content abundantly available on the internet. As such, they are taking on the role of curators of content, i.e. editors and commissioners of programmes and shows (Sørensen 2014). And in their scheduling and production strategies they are doing all they can to manage the audience and user flows between the different platforms, for example in creating, hosting and curating themed online portals on cooking and gastronomy. Likewise, as discussed in Chapters 2, 3, 4, 5 and 6, TV programmes are starting points around which you can do other things on other screens, like accessing extra programme-related content online or interacting with other people (Sørensen 2014). Hence, the connection between social media and centrally produced media, in essence built on contrasting paradigms of communication (i.e. horizontal versus vertical, multipoint-to-multipoint versus point-to-multipoint), is eagerly explored and increasingly exploited. The aim, in the words of Couldry (2009), is to reproduce the old idea that the media, and broadcasting in particular, is 'our privileged access point to society's centre or core' and 'that what's "going on" in the wider world is accessible first through a door marked "media"' (op. cit., 440). Although perhaps to young people the reference point par excellence that explains the social world's functioning is Facebook and YouTube, broadcasting companies are doing all they can to keep their pole position in the networked video culture.

This is not an easy task, because of the profoundly individualized engagement with audiovisual content through all kinds of on-demand delivery technologies, which is the outcome of the 'anytime, anywhere' paradigm of media use so dominant in our age. However, some are pointing out that 'liveness', one of the foundations of broadcasting discussed in Chapter 5, is also coming back in contemporary video and social media cultures. Technological experiments with user-generated live broadcasts are exploring the technological opportunities for co-producing and broadcasting live footage while being individually mobile (see Juhlin et al. 2014). Does this suggest that liveness, the phenomenological experience of being virtually connected with others at the time the event – banal as it might be in the case of, for example, Snapchat, or politically relevant, as in case of the Arab spring – takes place in a context far removed from traditional broadcasting?

Probably, sport programming represents one of the most interesting examples of how broadcasting is sustaining its traditional meaning as a central, shared cultural forum in a visual culture that is otherwise heavily dominated by other dispersed media players, such as YouTube, the official websites of FIFA and the national football teams, channel video-streaming websites, etc. Historically, sport as seen on TV – and for many countries in the world, football – is the symbolic space where feelings, ideas and thoughts about national community, ideals and values are constructed and negotiated. The interesting paradox about sport is that as an old, traditional broadcasting content – radio too has always paid a lot of attention to sport and still does – sport has at the same time easily and quickly become the quintessential element of digital broadcasting. We see this, as Johnson (2009) argues, in services making sport available through mobile phone alerts, video streaming and interactive websites, all applications that have capitalized on people's idiosyncratic interests in sport. But at the same time, sport has not at all lost its centralizing potential and 'remains the most visible, ritual site of "broadcast address", gathering the largest, multi-demographic audiences for shared, "water-cooler" television experiences' (ibid.: 132). Sport also continues to play a crucial role in the construction of ritual media ceremonies which large groups of people attend, such as the opening ceremony of the Olympic games (Cui 2013).

DEMOCRATIZATION AND PARTICIPATION

Broadcasting has from its start contributed to the democratization of the public sphere and the participation of very different people in relevant domains of social and cultural life, albeit not in the sense of simply allowing direct two-way-communication or allowing broad access to the production processes and production of symbolic content. Nevertheless, the technologizing process of broadcasting raises concerns about socially and technically disadvantaged groups who might miss the boat of digital broadcasting. Research keeps on indicating that old demographics, such as age and education levels, still shape the types of media use people develop in their daily life. A US study, for example, unsurprisingly, revealed that highly educated people and people who use computers at work – obviously, these groups often overlap – are far more acquainted with online media, like online video sites, digital video streaming and web news and sports. Whereas less-educated people and older people show a more old-fashioned television-oriented repertoire (i.e. watching entertainment programming, commercial breaks and programme promotions, news programmes and random channel surfing). Interestingly, this study also showed that income did not have a significant relationship with the types of media repertoire that are accessed (Taneja et al. 2012: 963). However, from a more global perspective, it has been argued elsewhere in this book (see Chapter 5) that issues of social inequality remain underexplored, owing to a heavily biased research agenda focusing on the industrialized countries of the world, and within these countries attention is mostly paid to the more trendsetting forms of practices, often the kind of media uses researchers themselves display.

CONCLUSION

As the different chapters of this book have tried to explicate, the evolution of broadcasting is not coming to an end, but rather is entangled in the circular interplay of four processes, which McLuhan (see McLuhan and McLuhan 1988) has coined as enhancement, obsolescence, retrieval and reversal (see Chapter 1). Throughout this book, we hope that we have made it clear that all four processes are being observed, described and empirically substantiated. For a start, many scholars and researchers have reached agreement on the idea that the new forms of broadcasting are in multifarious respects enhancing the way people can engage with televisual and radio culture. The booming of so-called 'video interaction' (Juhlin et al. 2014) is but one development that is believed to alter fundamentally consumer and producer practices in ways that are enhancing creativity, participation, diversity and liberation from institutional constraints. The explosion of video content online has enabled new trends such as collective and crowd-sourced video production where people load up their own video footage to contribute to larger documentaries, artistic creations like music video clips and current affairs programming. This epitomizes the 'increasingly flattened hierarchy between, on the one hand, what used to be a well-defined group of production professionals, and on the other hand, the masses of passive viewers of the same media' (Juhlin et al. 2014: 685). But this type of enhancement is still not mainstream, and the question remains as to which audiences exactly are empowered by all these exciting technological developments (Bird 2011).

In contemporary research, both empirical and theoretical, we find very few proponents claiming broadcasting will become obsolete. Yet, a number of scholars are delving into the problem of substitution and displacement, by pointing at the new audience practices of the (pre-eminently young) second-screen, social-media generations (Bondad-Brown et al. 2012; Cha and Chan-Olmsted 2012). These have grown up with the internet, MP3, mobile phone and tablet, they are abandoning traditional ways of watching television, they have drifted away from, in Couldry's terms as mentioned above, the social centres that broadcasting channels always have been, and on-demand and real-time access to audiovisual culture has become for them self-evident. They are often believed to be breaking down the vestiges of traditional broadcasting. From a macroscopic perspective, this raises intriguing questions about how these generations will define and construct their citizenship identities, and come to grip with the boundaries of the society, culture and political systems of which they are part.

However, as explained in the introduction of this book, McLuhan has drawn our attention to the fact that obsolescence does not simply lead to the disappearance of a medium, but rather restarts a new circular movement in which old meanings and characteristics of a medium are retrieved, albeit never in the same form and way as they used to be. That is why, in the 'morass of audiovisual content' (Sørensen 2014: 37) we are dealing with today, a lot of old-established programme concepts, genres and formats have been recovered from the past – like the old talent scouting shows of early television's past – but remediated in a style that appeals to twenty-first century audiences. Interesting in this context too is television's role on the internet in keeping individual memories alive and in constructing collective memories to which larger cultures of multifarious types (national, global, subcultural, diasporic, etc.) adhere. For example, quite a lot of YouTube's material is old TV and video clips of music bands from days long past (The Beatles, The Rolling Stones, The Doors, etc.), children series and films we used to watch, and great historical moments (the Kennedy assassination, the first man on the moon, the rise and fall of the Berlin wall, etc.). This kind of 'mediated historicity', in the words of Thompson (1995), has expanded with the advance of digital archiving technologies, but it is definitely television, and broadcasters in particular, that are catering to the nostalgic taste of audiences (Lozano 2013). McLuhan's 'retrieval' (see Chapter 1) is also shown in the way media industries are re-inventing their business models and structures without giving up the old basics of capitalist economy, i.e. competitiveness, capital accumulation, expansion and diversification. Especially the last process, as discussed in Chapter 3, in close connection with all the technological possibilities that are being developed, might be very influential in the reconfiguration of broadcasting's ontology, an idea that was definitely lacking from McLuhan's frame of analysis. One can imagine all forms of his notion of 'reversal' when thinking about broadcasting that is being pushed to the limits of its potential – from a mass medium to a niche medium, from a central medium to a marginal medium, from a popular medium to an unattractive medium, from a mundane, banal and ordinary medium to an unusual and original medium and from an inferior-quality medium to a high-quality medium. But what actual changes will take place will depend considerably upon the continuously renegotiated balance of power between the old and new industrial players in the mediascape.

Chapter Summary

■ In this concluding chapter we have attempted to summarize the key socio-cultural trends and political-economic developments in digital broadcasting by linking them up with five theoretical concepts: liquidization, defamiliarization, storytelling, de- and recentralization, and democratization and participation.

■ In line with the general aims of this book, this final chapter discussed both the newness and oldness of digital broadcasting.

■ Lastly, this chapter pointed out issues that are or should be given priority in research on the changes broadcasting is going through.

ANNOTATED GUIDE TO FURTHER READING

Chapter 1 Introduction

In the introductory chapter we discuss the place and meaning of digital broadcasting in the new media field. Lievrouw and Livingstone (2002) provide a helpful introduction to new media generally, which contains similar perspectives to the ones discussed in this book. Lister et al. (2003) also give an insightful account of new media (including digital broadcasting) from a media studies perspective.

When focusing on digital broadcasting one of the first anthologies is Spigel and Olsson's (2004), identifying the major issues for television in the period of transition. The essays in this edited book consider the future of television in the US and Europe and comprise a collection of historical, critical and speculative essays by television and media scholars. Another insightful discussion of the relevance of broadcasting – in particular television – in the digital age can be found in a 2009 special issue 'The End of Television? It's Impact on the World (So Far)' of *The ANNALS of the American Academy of Political and Social Science*, put together by Katz and Scannell. A good overview can be found in the introductory article by Katz (2009). Geiger and Lampinen (2014) give a more recent perspective in the *Journal of Broadcasting and Electronic Media*. This is also the introduction to their special issue 'Old Against New, or a Coming of Age? Broadcasting in an Era of Electronic Media'.

Similar ontological discussions of the notion of television in the digital age can be found in the edited volume by de Valck and Teurlings (2013). The goal of this book is to think through the implications of television's transformations for television theory today. Other edited volumes that provide multifaceted insights on television in a digital context are Gripsrud (2010a) and Bennett and Strange (2011). These works have a more European focus, while Kackman et al. (2011) deal more with the change from analogue to digital broadcasting in the US context. Turner and Tay (2009) also pay attention to local variations in the reconfiguration of television in the post-broadcast era. Other volumes in this area are Gerbarg's (2009) and Alvarez-Monzoncillo's (2011).

Some works do not particularly focus on digital television, but do discuss the meaning of and changes in television in more general ways. In this vein, Buonanno (2008) takes a look at the evolution of television studies, and discusses the influence of past, present and future television on society.

Chapter 2 A Historical Approach to Digital Broadcasting

Williams ([1974] 2003) has written a landmark study on the evolution of television broadcasting. He gives a historical account of the origin and evolution of (analogue) television, combining technological, economic, social and cultural perspectives. Some other classic monographs that propose a historical segmentation of television's development since 1950 are Ellis (2000) and Lotz (2009). The major subdivision used in this chapter is based on the three ages of television that Ellis (2000) has coined: scarcity, availability and plenty. In his book he discusses the transformation of television and its role in the modern world as part of the changing consumer culture of the twentieth century. Lotz (2009) is the much-cited first single-author book on the contemporary evolution on television, mainly from a US perspective. It is based on trade publications and interviews with key players. While the latter takes a more economic and political perspective, Meyrowitz (1985) and Spigel (1992) provide seminal social and cultural analyses of the evolution of television. Peters (2009) is helpful for understanding the more recent historical evolution of (new) media and broadcasting, by giving a structured overview of the historical development of and communication research on new media. McChesney (1993), Abramson (2003), Schwartz (2003) and Mullan (2008) make up some of the other relevant accounts of (parts of) the history of television, focusing mainly on the US.

Radio broadcasting has been underexposed in the literature in comparison to television. However Douglas (1989) describes the American history of radio between 1899 and 1922 as it passed through invention, innovation and regulation. This is followed by Douglas (1999), which covers the US public response to radio from the 1920s onwards. Smulyan (1994) gives a historical account of the commercialization of US radio broadcasting between World War I and World War II. Crisell (1994) looks at the role radio played in the development of modern popular culture.

Winston (1998) offers a more general historical account of radio and television technology, in addition to other communications technologies. The book argues that the development of these technologies is the product of a constant play-off between social necessity and suppression. Flichy and Libbrecht (1995) provide a comprehensive social history of various communication technologies since the end of the eighteenth century. The text focuses on the intimate relationship between technological, policy and social change. Michalis (2007) and Donders (2011) both give an overview of how communication policy has evolved in Europe, while Hitchens (2006) makes a detailed analysis of policy and regulatory measures relating to the promotion of media diversity in the UK, US and Australia.

Chapter 3 The Broadcasting Industry

In order to critically assess how the industry in general operates (on a macro level) – and thus also digital broadcasting – we refer to the political economy literature. Some key references for this field are the seminal book Mosco (2009), and the two edited volumes Wasko et al. (2011) and Winseck and Jin (2011). These publications discuss the political economy of media and communication, which refers to the broadcasting industry.

In contrast – on a micro level – there are many books discussing business economics in broadcasting and other media-related domains, like Griffiths (2003), Wicks et al. (2004) and Aris and Bughin (2005). They often start from management theories, referring to classic works like that of Porter (1980, 1985).

A very promising area for studying the broadcasting industry – on a meso level – can be found in the domain of media economics. A good starting point is the monograph by Doyle (2002), which provides an introduction to the key economic concepts and issues affecting the media sector. Two edited works that cover similar topics are Seabright and von Hagen (2007) and Küng et al. (2008). The latter cross-disciplinary book explores the implications of the internet for the media industries from economic, regulatory, strategic and organizational perspectives. Other relevant books in the area of media economics are Picard (2011), Hoskins et al. (2004) and Albarran (2002), as well as the edited handbook Albarran et al. (2005). Finally, Barwise and Picard (2012) take an economics and policy perspective on digital television.

Within the chapter we also discuss and apply value networks and business modelling as particular media economics concepts. Ballon (2007) is a good start for more background information on this approach. Evens and Donders (2013) are a good source in order to understand how to apply the latter to the broadcasting industry.

The technological details of digital television are extensively covered by Alencar (2009). In addition Lekakos et al. (2007) take a human–computer interaction perspective on interactive digital television applications, also discussing technological and advertising issues in relation to this. Several contributions in O'Neill et al. (2010) cover the technicalities of digital radio.

Finally, there are many books that focus on the policy and regulatory issues related to the digital broadcasting industry and the digital switchover, often combined with an analysis of the technological platforms. Starks (2013) gives an overview and analysis of the digital television switchover in the UK, Europe and worldwide. The book shows how lessons can be learned and transferred from one country to another, in order to inform the public debate about media policy during and after the switchover process. Galperin (2007) and the edited publication Cave and Nakamura (2006) take a more global perspective of digital television. Galperin (2007) discusses the digitization process and policy in UK and US from a political economy perspective, while Cave and Nakamura (2006) focus on policy questions, legal issues and technological platforms in US, Europe and Japan.

The edited books Van den Broeck and Pierson (2008) and Brown and Picard (2004) provide a detailed overview of the evolution of digital television and the situation in Europe, with special attention to the terrestrial variant. The monograph by Papathanassopoulos (2002) deals with the key issues surrounding digital television in Europe on policy and content levels and outlines possible future trends. Some publications like Debrett's (2010) and the edited work by Iosifidis (2010) focus specifically on public service broadcasting in Europe and discuss the implications of digitization and convergence.

Being one of the pioneers of digital broadcasting, the UK receives considerable attention in the literature. Starks (2007) studies UK policy regarding the digital switchover. It includes some comparative studies spanning the US, Japan and the leading countries of Western Europe. In Gardam and Levy (2008) a range of authors – from broadcasters, producers, politicians to academics – take a closer look at the notions of plurality, choice and diversity within UK public service broadcasting in the digital age.

Chapter 4 Production in the Digital Era

Compared to the literature dealing with the economic organization of the radio and television industry, far less research has been published on the social dimensions of working in the broadcasting industry. However, particularly since the 1970s, a strand has emerged within media sociology that has dealt with media professionals, production processes within the media industry, the social context of media production and the social organization of media work. Three particular spheres of media production have received attention from the very beginning: journalism, Hollywood studio production and television production. Notable scholars in this field are Tunstall (1971, 1993), Cantor ([1971] 1988) and Gitlin (1994). These works provide a good insight into the way that media professionals, working in different content branches, organize their daily work, build their careers, think about themselves and their audiences and struggle with creativity and autonomy within the economic and organizational constraints of media industries. Recent volumes on media sociology in which considerable attention is paid to contemporary issues related to media production are those of Schoemaker and Reese (2014) and Waisbord (2014).

With the growing importance of the creative and cultural industries, scholars working in this area have also contributed to the research on media professionals. Hesmondhalgh's edited work (2006), used in this chapter, discusses the work processes that lead to concrete media products. Various forms of television production (from journalism to animation) are presented within this book. His other publications show a particular interest in the power-related questions about creative labour, such as the tension between creativity and commercialization and the challenge of user-generated content for media professionals (see, for example, Hesmondhalgh 2010; Hesmondhalgh and Baker 2008, 2011).

This brings us to the process of the digitization of media and more particularly the impact of the internet and various types of do-it-yourself technologies on the work of traditional media professionals working in newsrooms, TV studios and on film sets. Obviously, this has renewed the interest in media production, and more particularly, in the way that media work is being reorganized and revised vis-à-vis the technological, economic, social and cultural changes to which digital media are contributing. In this respect, the work of Deuze (2007) gives a critical, empirically substantiated and topical overview of how digitization is shaking up the way in which media professionals work.

In addition to the academic publications that theoretically and empirically investigate processes of media production, there exists an extensive field of literature that describes the concrete procedures that are followed in media production processes. Rather than explaining, understanding and problematizing how the media work like they work, these handbooks give a descriptive overview of various TV and radio professions, their job content and related production skills and operational techniques. Here are but a few examples of a quite large number of handbooks in this area: Kindem and Musburger (2005), McLeish (2005), Owens and Millerson (2009) and Zettl (2012).

Chapter 5 Channels in the Digital Broadcasting Era

Within the European branch of media and communication studies, research on broadcasting channels has mainly focused on public service broadcasters. The UK has a particularly strong research tradition in this area. A substantial field of literature on the BBC, not only from British, but also non-British, scholars can be found both in books and academic journals. Other northern-European countries, such as Belgium (Flanders), the Netherlands and the Scandinavian countries – all countries with quite a strong public service tradition – also display a well-developed interest in public service broadcasting channels and their survival and renewal strategies in the digital era. For example Bardoel and d'Haenens (2008), Donders and Moe (2014), Syvertsen (2003) and Van den Bulck (2007, 2008) are illustrative of the northern-European interest in public service broadcasters. Within the international network RIPE (Re-Visionary Interpretations of the Public Enterprise), a non-profit organization of scholars and practitioners involved with the study, development and management of public service media organizations, interesting work is being published and presented on the contemporary meaning of public service broadcasting channels.

With the commercialization of the European radio and television landscape, the interest in channels' branding and marketing strategies has risen. Again, the public service broadcasters' branding and marketing efforts to stand up to the fierce competition from other channels is often analyzed, mainly from a critical perspective, as shown in, for example, Johnson (2012). Very little research can be found on the role and meaning of commercial broadcasting channels in European countries. If studied at all, the focus is mostly on comparative analyses (public service broadcasters vs. commercial broadcasters) of television programming and content. Yet, the role and meaning of these channels for European society and democracy remains underexplored. Interesting work in this field, however, can be found in Gripsrud and Weibull (2010) and Donders, Pauwels and Loisen (2013).

Obviously, in the US and Canadian broadcasting context, which from the start has developed differently from the European, much more attention has been paid to the commercial television networks and stations. The success of particular television channels, such as HBO for example, and the increasing competition in the television market has renewed an interest in issues of channel identity and branding – see, for example, the study of Ali (2012). A very interesting book in this context is again that of Johnson (2012). This author argues that branding has become the key strategy to respond to greater competition in the digital, multi-channel and multiplatform industry, both in the UK and the US. Critical discussions of the ideological power of the contemporary culture of promotion and marketing can be found in Caldwell's (2008) work. There is also the more marketing- and management-oriented tradition of research that investigates which branding strategies are most successful to improve the relationship with the audience in the crowded radio and television environment. Recent studies by Lis and Post (2013), and Peirce and Tang (2012) represent this area. A good overview of the research in this field can be found in Malmelin and Moisander (2014).

Last, there exists an expanding field of interesting literature on the new 'channels' of broadcasting, such as YouTube, Netflix, Hulu, etc. The edited volumes of Burgess and Green (2009b), Kavoori (2011) and Snickars and Vonderau (2009) all provide a good overview of how YouTube has established itself in economic, social and cultural terms. In particular the political-economical handbook edited by Wasko et al. (2011) presents a good critical discussion of the rise of these new players, and how they are part of the broader, persistent industrial systems of profit making.

Chapter 6 Audiences in the Digital Broadcasting Era

Audiences of digital broadcasting can be investigated from very different angles. McQuail (1997) gives a general overview of the different ways that audiences are framed. The edited volume of Nightingale (2011b) discusses the complexity of audience(s) and the research traditions that have developed over the past century. As we talk about digital broadcasting, both the 'digital' and the 'broadcasting' components are relevant in relation to the literature on audiences. The first part refers to audiences as users of digital technologies, while the second part relates to audiences as viewers of television and listeners of radio.

'Audiences as users' in a technological context can typically be found in the tradition of Science and Technology Studies (STS), which looks at the mutual shaping of technologies and society. MacKenzie and Wajcman (1999), Williams and Edge (1996) and Bijker et al. (1987) are some classic works in this area, while Whitworth and de Moor (2009), Hackett et al. (2008) and Flichy (2007) offer more recent updates and give a good overview of the diversity of the field. Gillespie et al. (2014) discuss how insights from STS and from media and communication studies can be integrated. Oudshoorn and Pinch (2003) is a classic edited volume that focuses on the role of the user in STS. The edited volumes by Pierson et al. (2008) and Pierson, Mante-Meijer and Loos (2011), based on the European research network COST 298 'Participation in the broadband society', provide an overview of how users have been investigated and involved in technological research and innovation, also outside the field of STS.

Other perspectives on audiences as users of technologies and innovations can be found in diffusionism and in domestication research. Diffusionism refers to the seminal work by Rogers (2003), where he describes the adoption and diffusion of innovations, often translated as technological innovations. The latter also has links with the Technology Acceptance Model (TAM), introduced by Davis (1989), looking at the perceived usefulness and the ease of use of technologies. As for domestication research, Silverstone and Hirsch (1992), Silverstone and Haddon (1996) and Haddon (2004) are key sources in order to understand how media technologies become 'domesticated' in everyday life. Berker et al. (2005) give an update of a decade of domestication research, while Haddon (2006) evaluates how this framework has contributed to understanding the usages of media and ICT.

'Audiences as viewers/listeners' has a long tradition in media and communication studies. It was Williams ([1974] 2003) who paved the way for two central academic traditions in the study of broadcasting audiences: cultural studies and political economy. The cultural studies tradition focused on the viewers and listeners of popular (media) culture, like soaps, entertainment show, news etc. Hall (1973) is perceived as a key text that introduces the notion of the 'encoding' and 'decoding' of these (media) texts. Ang (1996), Gauntlett and Hill (1999), Moores (1993), Morley (1986) and Fiske (1987) are seen as some of the major contributions to the study of television audiences. The political economy tradition did not originally pay specific attention to television and radio audiences. Its main goal was to understand how the broadcasting industry serves its own (class) interests with the production of mass media. This was pioneered by the Frankfurt School around the 1930s. Adorno and Horkheimer ([1944] 1986) has been followed by other key contributions, like Habermas ([1962] 1989) on the public sphere and Smythe (1977, 2006) on the audience commodity. The latter has been very influential in the critical study of commercial media. More recent critical perspectives on audiences and users in the digital age can be found in Bermejo (2009) and Fuchs (2014). Napoli (2010b) gives a clear insight into how the meaning of media audience continues to evolve in relation to changing technological, economic and political conditions.

Chapter 7 Rethinking Digital Broadcasting and New Media

The discussion of the transformation of television has emerged in several interesting volumes, all exploring both the continuities and innovations within television in the digital age (Bennett and Strange 2011; Gripsrud 2010a; Kackman et al 2011; de Valck and Teurlings 2013). The digitization of radio, albeit to a far lesser extent, has also been dealt with in various edited volumes. In Gazi et al. (2011), for example, the impact of new technologies on radio is highlighted from a European perspective. Dubber's book (2013), published in the *Digital Media and Society* series, provides a critical discussion on the changing meaning and definition of radio in a predominantly digital media environment.

For the most up-to-date developments in theoretical thinking about and empirical research into radio and television, academic journals such as *New Media and Society*, *Television and New Media*, *International Journal of Digital Television*, *Journal of Broadcasting and Electronic Media* (with, for example, its 2014 special issue 'Old Against New, or a Coming of Age? Broadcasting in an Era of Electronic Media') and *Convergence: the International Journal of Research into New Media Technologies* are devoted to the most recent trends in the study of television and broadcasting. The academic journal *Telematics and Informatics*, which takes an interdisciplinary and global approach to ICT, does not particularly focus on radio and television, but the digitization of these media and their increasing hybridization with the internet and mobile technologies, has increased the attention the journal gives to the topics that we have dealt with in this book. See, for example, its special issues on the impact of digital media on radio news culture, and also the economic and policy aspects of television distribution.

EXERCISES AND QUESTIONS

Chapter 1 Introduction

Exercises

1. Check the studies and statistics that exist in your own country concerning the time that people spend on watching television and listening to radio. How has this evolved over the past years? Would you conclude that television and radio are becoming obsolete? Why (or why not)?

2. Reflect on the role and meaning that television has in your everyday life, with the following questions as guidelines: How much time do you watch on average? Where do you watch? What is the division between watching alone and viewing television with other people? How many devices do you use? To what extent do you watch linear broadcasting in comparison to recorded, downloaded or streamed programmes? Do you spend money on television programmes (e.g. pay-TV, video-on-demand)? Then compare this with how other people experience broadcasting nowadays: your peers, older people (e.g. parents) and younger people. What are the most striking similarities and differences?

Questions

1. Why is digital broadcasting a good example of 'media renewability', framed as a five-stage process (Chapter 1), in terms of: (1) technical invention; (2) cultural innovation; (3) legal regulation; (4) economic distribution; and (5) social mainstream?

2. In what ways do digital and converged television technologies shape government policies and industry strategies, and in what ways are they being shaped by those policies and strategies? What are the possible opportunities and threats for citizens?

3. Will the traditional division in broadcasting between media professionals (who make the media) and media amateurs (who consume the media on offer) gradually disappear, owing to the fusion of broadcasting and the internet, and the co-creative potential of user-generated content?

4. Tim Cook (CEO Apple) claimed in a 2014 interview that television is still stuck in the 1970s. Do you agree? Why (of why not)? (www.youtube.com/watch?v=oBMo8Oz9jsQ)

Chapter 2 A Historical Approach to Digital Broadcasting

Exercises

1. Think about your grandparents' home and your own home when you were young. How were the radio and TV sets embedded in these homes, in terms of their location and how they were used? And how has this changed if you consider their homes today?
2. Generate a list of other technologies that, together with media and communication technologies, have contributed to the construction of the home theatre, mobile home and smart home.
3. As a group, imagine an alternative history of broadcasting, i.e. different from the history as we know it, and reflect on how broadcasting might have looked today.

Questions

1. Critically discuss the main differences between the European and US history of broadcasting policy.
2. Critically discuss the relationship between Williams' notion of 'mobile privatization' and Spigel's notion of 'privatized mobility'.

Chapter 3 The Broadcasting Industry

Exercises

1. This exercise will take the form of a class debate between different groups. Each group of students represents a stakeholder in the broadcasting industry: content producer, content aggregator, distributor and consumer. Organize a discussion in which each group argues why they are the most suitable actor to take the lead in broadcasting in the digital era.
2. Imagine that you would like to start a new business in the digital broadcasting industry. Which (possibly new) role in the changing business model do you think would have the best prospects? Why?

Questions

1. Explain the five main business roles in the digital broadcasting value network: (1) content creation and production; (2) content aggregation and packaging; (3) content distribution; (4) content service provision; and (5) content service and technology consumption. Give an example of a stakeholder of each of these roles in your own country. Discuss why this type of horizontal perspective

on digital broadcasting is more helpful than differentiating vertical sectors (television, the internet, telecommunication, software, etc.).

2. Re-examine the figure of the generic value network for (digital) television and distribution. In which of the business roles could the user play a (significant) role?

3. How would you define radio in the current digital media industry? In what ways is radio a broadcast medium of the past, and/or to what extent can it remain a medium of the future?

4. Situate the importance of 'customer ownership' in a value network, from the perspective of marketing, billing and customer intelligence (via data and metadata).

Chapter 4 Production in the Digital Era

Exercises

1. Imagine that you are in a production team developing a new television programme format that is not part of the larger reality TV-genre. In what ways could you try to incorporate fan- and amateur-based content into your programme format?

2. As a group, brainstorm to find any examples where ordinary people have become popular without the intervention of the media industries?

3. Make a list of indispensable ingredients for scoring a transmedia storytelling-hit.

Questions

1. To what degree have economic, technological and cultural developments influenced the popularity of reality TV? Explain how these three developments interrelate.

2. Given Jenkins' (2006) ideas about convergence culture, discuss whether there are new television programmes, genres or formats conceivable that are either consumer-driven or industry-driven.

Chapter 5 Channels in the Digital Broadcasting Era

Exercises

1. As a group, can you say, for each television and radio programme that you like to watch or listen to, on which channel they are broadcasted?

2. How would you research the claim that watching TV in a non-linear, on-demand way is threatening the traditional role that television content has always played in everyday conversations and community life?

3. Think about examples of how channels in your country are rebranding themselves in response to the (growing) success of online video delivery services (like, for example, Netflix)?

Questions

1. Does the notion of broadcasting only apply to television channels and radio stations, or can you name other distributors that can be considered as broadcasters? Critically discuss with reference to examples.
2. Critically discuss the pros and cons of the different channel ages (from channel scarcity to channel abundance) for the audience.
3. Critically evaluate the main arguments we hear in the debate about the role of channels for democracy and social life, by providing some counter-arguments that refute this line of reasoning.

Chapter 6 Audiences in the Digital Broadcasting Era

Exercises

1. In a group, discuss how you as an audience member experience different types of content (movies, news, live sport, talent shows, etc.) on different media technologies (television screen, computer screen, tablet, smartphone, etc.). What kind of similarities and differences do you observe (time spent, possibility for 'binge viewing', degree of interaction, togetherness, immersion, etc.)?
2. Set up an experiment with the class where each student or group of students watches television for a whole week via only one type of broadcasting (analogue television via antenna with no recording option, digital television with a set-top box for recording, television via the computer and the internet (e.g. Netflix, Hulu), channel app on a tablet, live streaming on a smartphone via wireless internet, etc.) without using other access devices for television that week. Then share your experiences. Did you miss something? Did you feel imprisoned by having a limited choice of device? Did you concentrate more or less while viewing television? Did you watch other types of programmes from the ones that you normally watch? Did you watch more alone or more with other people? To what extent were your viewing habits different from your regular ones?

Questions

1. 'How we understand ourselves and build our identities, how we are able to take advantage of the world around us, is intrinsically shaped by our audience affiliations and practices.' (Nightingale 2011a: 1) Would you agree with Nightingale? Why (not)?
2. Which kind of (new) affordances does digital broadcasting have for each of the four functional domains in the household: (1) entertainment; (2) news; (3) organization of daily life; and (4) sociality? To what extent do you and your household make use of these affordances in your everyday life?

3. Describe if and to what extent you see the 'appointment model' of traditional broadcasting media as still being relevant in the current digital media landscape. Are there still occasions when specific kinds of programmes and/or their timing influence your TV evening, possibly together with family or friends? Or are we all moving towards a 'networked video culture'?

Chapter 7 Rethinking Digital Broadcasting and New Media

Exercises

1. As a group, brainstorm to find any examples of how new forms of broadcasting might contribute to a more democratic organization of political, social and cultural life.
2. How would you research the multiplatform use of broadcasting content? Can you draw up a research proposal that would enable you to observe and understand people's engagements with audiovisual content?
3. From what you have learnt from this book, which aspects of broadcasting might become obsolete in the future?

Questions

1. Drawing upon McLuhan's four laws of media change, and in the light of developments in today's broadcasting, give some other examples of so-called 'retrieval'?
2. Why is it useful to study broadcasting from a multifaceted approach?

BIBLIOGRAPHY

Aaker, D. A. (1991), *Managing Brand Equity: Capitalizing on the Value of a Brand Name*, New York: The Free Press.

Aarts, E. and Marzano, S. (eds) (2003), *The New Everyday: Visions of Ambient Intelligence*, Rotterdam: 010 Publishers.

Abramson, A. (2003), *The History of Television, 1942 to 2000*, Jefferson, NC: McFarland.

Adorno, T. and Horkheimer, M. ([1944] 1986), *Dialectic of Enlightenment*, London: Verso.

Ala-Fossi, M. (2010), 'The technological landscape of radio', in B. O'Neill, M. Ala-Fossi, P. Jauert, S. Lax, L. Nyre, and H. Shaw (eds) *Digital Radio in Europe: Technologies, Industries and Cultures*, Bristol: Intellect, 43–65.

Albarran, A. (2002), *Media Economics: Understanding Markets, Industries and Concepts* (2nd edn), Ames: Iowa State University Press.

Albarran, A., Chan-Olmsted, S. and Wirth, M. (eds) (2005), *Handbook of Media Economics and Management*, Mahwah, NJ: Lawrence Erlbaum Associates.

Albarran, A. B., Anderson, T., Bejar, L. G., Bussart, A. L., Daggett, E., Gibson, S., Gorman, M., Greer, D., Guo, M., Horst, J. L., Khalaf, T., Lay, J. P., McCracken, M., Mott, B. and Way, H. (2007), '"What happened to our audience?" Radio and new technology uses and gratifications among young adult users', *Journal of Radio Studies*, Vol. 14, No. 2: 92–101.

Alencar, M. S. (2009), *Digital Television Systems*, Cambridge: Cambridge University Press.

Ali, C. (2012), 'Of logos, owners, and cultural intermediaries: defining an elite discourse in re-branding practices at three private Canadian television stations', *Canadian Journal of Communication*, Vol. 37, No. 2: 259–79.

Alvarez-Monzoncillo, J. M. (2011), *Watching the Internet: The Future of TV?*, Oporto: Media XXI Formalpress.

Anderson, C. (2006), *The Long Tail: Why the Future of Business is Selling Less of More*, New York: Hyperion.

—(2009), *Free: The Future of a Radical Price*, New York: Hyperion.

Andjelic, A. (2008), 'Transformations in the media industry: Customization and branding as strategic choices for media firms', in C. Dal Zotto and H. van Kranenburg (eds) *Management and Innovation in the Media Industry*, Cheltenham: Edward Elgar Publishing Limited, 109–29.

Andrejevic, M. (2002), 'The kinder, gentler gaze of *Big Brother*: reality TV in the era of digital capitalism', *New Media and Society*, Vol. 4, No. 2: 251–70.

—(2004), *Reality TV: The Work of Being Watched*, New York: Rowman and Littlefield.

Ang, I. (1996), *Living Room Wars: Rethinking Media Audiences for a Postmodern World*, London: Routledge.

Aris, A. and Bughin, J. (2005), *Managing Media Companies: Harnessing Creative Value*, Chichester: John Wiley & Sons, Inc.

Awan, F. (2007), *Young People, Identity and the Media: A Study of Conceptions of Self-Identity Among Youth in Southern England*, PhD thesis, Bournemouth University, Bournemouth, UK. Avalailable from http://eprints.bournemouth.ac.uk/10466/1/Fatimah_Awan.pdf

Bachmayer, S., Lugmayr, A. and Kotsis, G. (2010), 'Convergence of collaborative web approaches and interactive TV program formats', *International Journal of Web Information Systems*, Vol. 6, No. 1: 74–94.

Bain and Company (2007), *The Digital Video Consumer: Transforming the European Video Content Market*, s.l.: Liberty Global Policy Series.

Ballon, P. (2005), *Best Practice in Business Modelling for ICT Services* (TNO-ICT Report), Delft: TNO.

—(2007), 'Business modelling revisited: the configuration of control and value', *Info – The Journal of Policy, Regulation and Strategy for Telecommunications, Information and Media*, Vol. 9, No. 5: 6–19.

Bardoel, J. and d'Haenens, L. (2008), 'Public service broadcasting in converging media modalities: practices and reflections from the Netherlands', *Convergence: The International Journal of Research into New Media Technologies*, Vol. 14, No. 3: 351–60.

Barkhuus, L. and Brown, B. (2009), 'Unpacking the television: user practices around a changing technology', *Transactions on Computer–Human Interaction*, Vol. 16, No. 3: 11–22.

Barwise, P. and Picard, R. G. (2012), *The Economics of Television in a Digital World: What Economics Tells Us for Future Policy Debates*, Oxford: Reuters Institute for the Study of Journalism, Oxford University.

Bauman, Z. (2000), *Liquid Modernity*, Cambridge: Polity Press.

Bauwens, J. (2002), *Burgers voor de buis: een kwantitatief en kwalitatief publieksonderzoek naar de relatie tussen tv-consumptie en burgerschap*, PhD thesis, Vrije Universiteit Brussel, Brussels, Belgium.

—(2007), 'De openbare televisie en haar kijkers: oude liefde roest niet?', in A. Dhoest and H. Van den Bulck (eds) *Publieke televisie in Vlaanderen: een geschiedenis*. Gent: Academia Press, 91–124.

Beghetto, R. A. and Kaufman, J. C. (2007), 'Toward a broader conception of creativity: a case for "mini-c" creativity', *Psychology of Aesthetics, Creativity and the Arts*, Vol. 1, No. 2: 73–9.

Bennett, J. (2008), 'Television studies goes digital', *Cinema Journal*, Vol. 47, No. 3: 158–65.

—(2011), 'Introduction: Television as Digital Media', in J. Bennett and N. Strange (eds) *Television as Digital Media*, Durham: Duke University Press, 1–30.

Bennett, J. and Strange, N. (eds) (2011), *Television as Digital Media*, Durham: Duke University Press.

Berker, T., Hartmann, M., Punie, Y. and Ward, K. (2005), *Domestication of Media and Technology*, Berkshire: Open University Press.

Bermejo, F. (2009), 'Audience manufacture in historical perspective: from broadcasting to Google', *New Media and Society*, Vol. 11, No. 1/2: 133–54.

Berners-Lee, T. (1999), *Weaving the Web*, London: Orion Business Books.

Berry, R. (2003), 'Speech radio in the digital age', in A. Crisell (ed.) *More Than a Music Box: Radio Cultures and Communities in a Multi-Media World*, New York: Berghahn Books, 283–302.

—(2006), 'Will the iPod kill the radio star? Profiling podcasting as radio', *Convergence: The International Journal of Research into New Media Technologies*, Vol. 12, No. 2: 143–62.

Bignell, J. (2004), *An Introduction to Television Studies*, Abingdon: Routledge.

—(2013), *An Introduction to Television Studies* (3rd edn), Abingdon: Routledge.

Bijker, W. E., Hughes, T. P. and Pinch, T. J. (1987), *The Social Construction of Technological Systems: New Directions in the Sociology and History of Technology*. Cambridge, MA: The MIT Press.

Biltereyst, D. (2004), 'Public service broadcasting, popular entertainment and the construction of trust', *Cultural Studies*, Vol. 7, No. 3: 341–62.

BIPE (2002), *Digital Switchover in Broadcasting: A BIPE Consulting Study for the European Commission* (final report), Brussels: European Commission – DG Information Society.

Bird, E. S. (2011), 'Are we all produsers now?', *Cultural Studies*, Vol. 25, No. 4: 502–16.

Blumler, J. G. and Katz, E. (eds) (1974), *The Uses of Mass Communications: Current Perspectives in Gratifications Research*, London: Faber.

Boddy, W. (2002), 'New media as old media: Television', in D. Harries (ed.) *The New Media Book*, London: BFI, 242–53.

—(2003), 'Redefining the home screen: Technological convergence as trauma and business plan', in D. Thorburn and H. Jenkins (eds) *Rethinking Media Change: The Aesthetics of Transition*, Cambridge, MA: The MIT Press, 191–200.

Bolin, G. (2011), *Value and the Media: Cultural Production and Consumption in the Digital Markets*, Farnham: Ashgate.

Bolter, J. D. and Grusin, R. (2000), *Remediation: Understanding New Media*, Cambridge, MA: The MIT Press.

Bondad-Brown, B. A., Rice, R. E. and Pearce, K. E. (2012), 'Influences on TV viewing and online user-shared video use: demographics, generations, contextual age, media use, motivations, and audience activity', *Journal of Broadcasting and Electronic Media*, Vol. 56, No. 4: 471–93.

Bondebjerg, I. (2010), 'A new space for democracy? Online media, factual genres and the transformation of traditional mass media', in J. Gripsrud (ed.) *Relocating Television: Television in the Digital Context*, Abingdon: Routledge, 113–24.

Born, G. (2005), *Uncertain Vision: Birt, Dyke and the Reinvention of the BBC*, London: Vintage.

Bourdieu, P. (1983), 'The field of cultural production, or: the economic world reversed', *Poetics*, Vol. 12, No. 4–5: 311–56.

Bourdon, J. (2000), 'Live television is still alive: on television as an unfulfilled promise', *Media, Culture and Society*, Vol. 22, No. 5: 531–56.

Brown, A. and Picard, R. G. (2004), *Digital Terrestrial Television in Europe*, Mahwah, NJ: Lawrence Erlbaum Associates.

Bruns, A. (2008), *Blogs, Wikipedia, Second life, and Beyond: From Production to Produsage*, New York: Peter Lang.

Bull, M. (2004), 'Automobility and the power of sound', *Theory, Culture and Society*, Vol. 21, No. 4–5: 243–59.

—(2006), 'Investigating the culture of mobile listening: From Walkman to iPod', in K. O'Hara and B. Brown (eds) *Consuming Music Together: Social and Collaborative Aspects of Music Technologies*, Dordrecht: Springer, 131–50.

—(2009), 'The Auditory Nostalgia of iPod Culture', in K. Bijsterveld and J. van Dijck (eds) *Sound Souvenirs: Audio Technologies, Memory and Cultural Practices*, Amsterdam: Amsterdam University Press, 83–93.

Buonanno, M. (2008), *The Age of Television: Experiences and Theories*, Bristol: Intellect.

Burgelman, J.-C. (1989), 'Political parties and their impact on public service broadcasting in Belgium: elements from a political-sociological approach', *Media, Culture and Society*, Vol. 11, No. 2: 167–93.

Burgess, J. (2006), 'Hearing ordinary voices: cultural studies, vernacular creativity and digital storytelling', *Continuum: Journal of Media and Cultural Studies*, Vol. 20, No. 2: 201–14.

Burgess, J. and Green, J. (2009a), 'The entrepreneurial vlogger: Participatory culture beyond the professional-amateur divide', in P. Snickars and P. Vonderau (eds) *The YouTube Reader*, Stockholm: National Library of Sweden & Wallflower Press, 89–107.

—(2009b), *YouTube: Online Video and Participatory Culture*, Cambridge: Polity Press.

Burke, D. (ed.) (1999), *Spy TV: Just Who is the Digital TV Revolution Overthrowing?* Hove: Slab-O-Concrete Publications.

Burton, G. (2000), *Talking Television: An Introduction to the Study of Television*, London: Arnold.

Caldwell, J. T. (2004), 'Convergence television: Aggregating form and repurposing content in the culture of conglomeration', in L. Spigel and J. Olsson (eds) *Television after TV: Essays on a Medium in Transition*, Durham: Duke University Press, 41–74.

—(2006), 'Critical industrial practice: branding, repurposing, and the migratory patterns of industrial texts', *Television and New Media*, Vol. 7, No. 2: 99–134.

—(2008), *Production Culture: Industrial Reflexivity and Critical Practice in Film and Television*, Durham, NC: Duke University Press.

Cantor, M. G. ([1971] 1988), *The Hollywood TV Producer: His Work and His Audience*, New Brunswick: Transaction, Inc.

Cauberghe, V. and De Pelsmacker, P. (2006), 'Opportunities and thresholds for advertising on interactive digital TV: a view from advertising professionals', *Journal of Interactive Advertising*, Vol. 7, No. 1: 25–40.

Cave, M. and Nakamura, K. (eds) (2006), *Digital Broadcasting: Policy and Practice in the Americas, Europe and Japan*, Cheltenham: Edward Elgar Publishing.

Cawson, A., Haddon, L. and Miles, I. (1995), *The Shape of Things to Consume: Bringing Information Technology into the Home*, London: Avebury.

Cha, J. and Chan-Olmsted, S. M. (2012), 'Substitutability between online video platforms and television', *Journalism and Mass Communication Quarterly*, Vol. 89, No. 2: 261–78.

Chamberlain, D. (2010), 'Television interfaces', *Journal of Popular Film and Television*, Vol. 38, No. 2: 84–8.

—(2011), 'Scripted spaces: Television interfaces and the non-places of asynchronous entertainment', in J. Bennett and N. Strange (eds) *Television as Digital Media*, Durham, NC: Duke University Press, 230–54.

Chan-Olmsted, S. M. and Kim, Y. (2001), 'Perceptions of branding among television station managers: an exploratory analysis', *Journal of Broadcasting and Electronic Media*, Vol. 45, No. 1: 75–91.

Coe, L. (1996), *Wireless Radio: A Brief History*, Jefferson: McFarland & Company, Inc.

Consumerlab – Ericsson (2013), *TV and Media: Identifying the Needs of Tomorrow's Video Consumers. An Ericsson Consumer Insight Summary Report* (August 2013), Stockholm: Ericsson.

Cooke, L. (2005), 'A visual convergence of print, television, and the internet: charting 40 years of design change in news presentation', *New Media and Society*, Vol. 7, No. 1: 22–46.

Cordeiro, P. (2012), 'Radio becoming r@dio: convergence, interactivity and broadcasting trends in perspective', *Participations: Journal of Audience and Reception Studies*, Vol. 9, No. 2: 492–510.

Costera-Meijer, I. (2004), *De toekomst van het nieuws: hoe kunnen journalisten en programmamakers tegemoetkomen aan de wensen en verlangens van tieners en twintigers op het gebied van onafhankelijke en pluriforme informatievoorziening?*, Amsterdam: Otto Cramwinckel Uitgever.

Couldry, N. (2000), *The Place of Media Power: Pilgrims and Witnesses of the Media Age*, London: Routledge.

—(2004a), 'Liveness, "reality", and the mediated habitus from television to the mobile phone', *Communication Review*, Vol. 7, No. 4: 353–62.

—(2004b), 'Theorising media as practice', *Social Semiotics*, Vol. 14, No. 2: 115–32.

—(2009), 'Does "the media" have a future?', *European Journal of Communication*, Vol. 24, No. 4: 437–49.

—(2011), 'The necessary future of the audience … and how to research it', in V. Nightingale (ed.) *The Handbook of Media Audiences*, Malden: Wiley-Blackwell, 213–29.

—(2012), *Media, Society, World: Social Theory and Digital Media Practice*, Cambridge: Polity Press.

Crisell, A. (1994), *Understanding Radio*, London: Routledge.

Cui, X. (2013), 'Media events are still alive: the opening ceremony of the Beijing Olympics as a media ritual', *International Journal of Communication*, Vol. 7: 1220–35.

Curtin, M. (2009), 'Matrix media', in G. Turner and J. Tay (eds) *Television Studies after TV: Understanding Television in the Post-Broadcast Era*, Abingdon: Routledge, 9–19.

Dahlgren, P. (1995), *Television and the Public Sphere: Citizenship, Democracy and the Media*, London: Saga.

Davis, F. D. (1989), 'Perceived usefulness, perceived ease of use, and user acceptance of information technology', *MIS Quarterly*, Vol. 13, No. 3: 319–41.

Dawson, M. (2007), 'Little players, big shows: format, narration, and style on television's new smaller screens', *Convergence: The International Journal of Research into New Media Technologies*, Vol. 13, No. 3: 231–50.

—(2010), 'Television between analog and digital', *Journal of Popular Film and Television*, Vol. 38, No. 2: 95–101.

Dayan, D. (2009), 'Sharing and showing: television as monstration', *The ANNALS of the American Academy of Political and Social Science*, Vol. 625, No. 1: 19–31.

Debrett, M. (2010), *Reinventing Public Service Television for the Digital Future*, Bristol: Intellect.

De Dobbelaer, R., Paulussen, S. and Maescele, P. (2013), 'Sociale media en oude routines', *Tijdschrift voor Communicatiewetenschap*, Vol. 41, No. 3: 265–79.

De Moor, S., Schuurman, D. and De Marez, L. (2013), *Digimeter 2013: Adoption and Usage of Media and ICT in Flanders*, Gent: iMinds iLab.o.

de Valck, M. and Teurlings, J. (eds) (2013), *After the Break: Television Theory Today*, Amsterdam: Amsterdam University Press.

Deuze, M. (2007), *Media Work*, Cambridge: Polity Press.

Donders, K. (2010), *Under Pressure? An Analysis of the Impact of European State Aid Policy on Public Service Broadcasting: an explorative and future-oriented analysis of Flanders, Denmark and the United States Public Service Broadcasting. Marginalization or Revival as Public Service Media?*, PhD thesis, Vrije Universiteit Brussel, Brussels, Belgium.

—(2011), *Public Service Media and Policy in Europe*, Basingstoke: Palgrave Macmillan.

Donders, K. and Evens, T. (2011), *Broadcasting and its Distribution in Flanders, Denmark and the United States: An Explorative and Future-oriented Analysis* (A research report for SBS Belgium), Brussels: SMIT-MICT-IBBT.

Donders, K. and Moe, H. (2014), 'European State-Aid Control and PSB: Competition Policy Clashing or Matching with Public Interest Objectives?', in K. Donders, C. Pauwels and J. Loisen (eds) *The Palgrave Handbook of European Media Policy*, Basingstoke: Palgrave Macmillan, 426–41.

Donders, K. and Pauwels, C. (2008), 'Does EU policy challenge the digital future of public service broadcasting? An analysis of the Commission's state aid approach to digitization and the public service remit of public broadcasting organizations', *Convergence: The International Journal of Research into New Media Technologies*, Vol. 14, No. 3: 295–311.

Donders, K., Pauwels, C. and Loisen, J. (eds) (2013), *Private Television in Western Europe: Content, Markets, Policies*, Basingstoke: Palgrave Macmillan.

Douglas, S. J. (1989), *Inventing American Broadcasting, 1899–1922*. Baltimore, MD: Johns Hopkins University Press.

—(1999), *Listening In: Radio and The American Imagination*, Minneapolis: University of Minnesota Press.

Dourish, P. (2001), *Where the Action Is: The Foundations of Embodied Interaction*, Cambridge, MA: The MIT Press.

Dourish, P. and Bell, G. (2011), *Divining a Digital Future: Mess and Mythology in Ubiquitous Computing*, Cambridge, MA: The MIT Press.

Dovey, J. (2004) 'Camcorder cults', R. C. Allen and A. Hill (eds) *The Television Studies Reader*, London: Routledge, 557–68.

—(2008), 'Dinosaurs and butterflies – media practice research in new media ecologies', *Journal of Media Practice*, Vol. 9, No. 3: 243–56.

Dovey, J. and Rose, M. (2012), 'We're happy and we know it: documentary, data, montage', *Studies in Documentary Film*, Vol. 6, No. 2: 159–73.

Doyle, G. (2002), *Understanding Media Economics*, London: Sage.

—(2010), 'From television to multi-platform: less from more or more from less?', *Convergence: the International Journal of Research into New Media Technologies*, Vol. 16, No. 4: 1–19.

Druick, Z. (2009), 'Dialogic absurdity: TV news parody as a critique of genre', *Television and New Media*, Vol. 10, No. 3: 294–308.

Dubber, A. (2013), *Radio in the Digital Age*, Cambridge: Polity Press.

Ducheneaut, N. M., Oehlberg, L., Moore, R. J., Thornton, J. D. and Nickell, E. (2008), 'Social TV: designing for distributed, sociable television viewing', *International Journal of Human–Computer Interaction*, Vol. 24, No. 2: 136–54.

Edwards, L. H. (2012), 'Transmedia storytelling, corporate synergy, and audience expression', *Global Media Journal*, Vol. 12, No. 20: 1–12.

Elliott, A. and Urry, J. (2010), *Mobile Lives*, London: Routledge.

Ellis, J. (1982), *Visible Fictions: Cinema, Television, Video*, London: Routledge & Kegan Paul.

—(2000), *Seeing Things: Television in the Age of Uncertainty*, London: I. B. Tauris.

—(2007), *TV FAQ: Uncommon Answers to Common Questions about Television*, London: I. B. Tauris.

eMarketer (2010), 'Online video viewing shifts to long-form content', eMarketer – Digital Intelligence. Available from http://www.emarketer.com/Article.aspx?R=1007745

Enli, G. S. (2009), 'Mass communication tapping into participatory culture: exploring Strictly Come Dancing and Britain's Got Talent', *European Journal of Communication*, Vol. 24, No. 4: 481–93.

Esser, A. (2010), 'Television formats: primetime staple, global market', *Popular Communication*, Vol. 8, No. 4: 273–92.

Evens, T. and Donders, K. (2013), 'Broadcast market structures and retransmission payments: a European perspective', *Media, Culture and Society*, Vol. 35, No. 4: 415–32.

Evens, T., De Marez, L., Hauttekeete, L., Biltereyst, D., Mannens, E. and Van de Walle, R. (2010), 'Attracting the un-served audience: the sustainability of long tail based business models for cultural television content', *New Media and Society*, Vol. 10, No. 6: 1005–24.

Fiske, J. (1987), *Television Culture*, London: Routledge.

Flichy, P. (2007), *Understanding Technological Innovation: A Socio-technical Approach*, Cheltenham: Edward Elgar.

Flichy, P. and Libbrecht, L. (1995), *Dynamics of Modern Communication: The Shaping and Impact of New Communication Technologies*, London: Sage.

Fowles, J. (1992), *Why Viewers Watch: Reappraisal of Television's Effects*, London: Sage.

Fuchs, C. (2014), 'Dallas Smythe reloaded: Critical media and communication studies today', in L. McGuigan and V. Manzerolle (eds) *The Audience Commodity in a Digital Age: Revisiting Critical Theory of Commercial Media*, New York: Peter Lang.

Fusco, S. and Perotta, M. (2008), 'Rethinking the format as a theoretical object in the age of media convergence', *Observatorio (OBS*) Journal*, Vol. 7, No. 4: 89–102.

Galperin, H. (2007), *New Television, Old Politics: The Transition to Digital TV in the United States and Britain*, Cambridge: Cambridge University Press.

Gardam, T. and Levy, D. (eds) (2008), *The Price of Plurality: Choice, Diversity and Broadcasting Institutions in the Digital Age*, Oxford: Reuters Institute for the Study of Journalism.

Gates, B. (1996), 'Content is King'. Available from http://www.craigbailey.net/content-is-king-by-bill-gates/ (1 March 1996).

Gauntlett, D. and Hill, A. (1999), *TV Living: Television, Culture and Everyday Life*, London: Routledge.

Gazi, A., Starkey, G. and Jedrzejewski, S. (eds) (2011), *Radio Content in the Digital Age: The Evolution of a Sound Medium*, Bristol: Intellect.

Geerts, D. (2009), *Sociability Heuristics for Interactive TV: Supporting the Social Uses of Television*, PhD thesis, Katholieke Universiteit Leuven, Leuven, Belgium.

Geiger, R. S. and Lampinen, A. (2014), 'Old against new, or a coming of age? Broadcasting in an era of electronic media', *Journal of Broadcasting and Electronic Media*, Vol. 58, No. 3: 333–41.

Gerbarg, D. (ed.) (2009), *Television Goes Digital*, New York: Springer.

Gershenfeld, N., Krikorian, R. and Cohen, D. (2004), 'The internet of things', *Scientific American*, Vol. 291, No. 4: 76–81.

Gibson, J. J. (1977), 'The Theory of Affordances', in R. E. Shaw and J. Bransford (eds) *Perceiving, Acting and Knowing: Toward an Ecological Psychology*, London: John Wiley, 67–82.

Giddings, S. and Kennedy, H. W. (2010), '"Incremental speed increases excitement": bodies, space, movement, and televisual change', *Television and New Media*, Vol. 11, No. 3: 163–79.

Gillan, J. (2011), *Television and New Media: Must-click TV*, New York: Routledge.

Gillespie, M. (1995), *Television, Ethnicity and Cultural Change*, London: Routledge.

Gillespie, R. (2012), 'The art of criticism in the age of interactive technology: critics, participatory culture, and the avant-garde', *International Journal of Communication*, Vol. 6: 56–75.

Gillespie, T., Boczkowski, P. J. and Foot, K. A. (eds) (2014), *Media Technologies: Essays on Communication, Materiality, and Society*. Cambridge, MA: The MIT Press.

Gitlin, T. (1994), *Inside Prime Time*, London: Routledge.

Goggin, G. (2011), 'Going Mobile', in V. Nightingale (ed.) *The Handbook of Media Audiences*, Malden: Wiley-Blackwell, 128–46.

Golding, P. (2010), 'The cost of citizenship in the digital age: On being informed and the commodification of the public sphere', in J. Gripsrud (ed.) *Relocating Television: Television in the Digital Context*, Abingdon: Routledge, 207–24.

Grainge, P. (2010), 'Elvis sings for the BBC: broadcast branding and digital media design', *Media, Culture and Society*, Vol. 32, No. 1: 45–61.

Gray, F. (2003), 'Fireside issues: Audience, listener, soundscape', in A. Crisell (ed.) *More Than a Music Box: Radio Cultures and Communities in a Multi-Media World*, New York: Berghahn Books, 247–64.

Gray, J. (2010), '"Coming up next": promos in the future of television and television studies', *Journal of Popular Film and Television*, Vol. 38, No. 2: 54–7.

Greer, C. F. and Ferguson, D. A. (2011), 'Using Twitter for promotion and branding: a content analysis of local television Twitter sites', *Journal of Broadcasting and Electronic Media*, Vol. 55, No. 2: 198–214.

Griffiths, A. (2003), *Digital Television Strategies: Business Challenges and Opportunities*, New York: Palgrave Macmillan.

Gripsrud, J. (ed.) (2010a), *Relocating Television: Television in the Digital Context*, Abingdon: Routledge.

—(2010b), 'Preface', in J. Gripsrud (ed.) *Relocating Television: Television in the digital context*, Abingdon: Routledge, xv–xxi.

—(2010c), 'Television in the digital public sphere', in J. Gripsrud (ed.) *Relocating Television: Television in the digital context*, Abingdon: Routledge, 3–26.

Gripsrud, J. and Weibull, L. (eds) (2010), *Media, Market & Public Spheres: European Media at the Crossroads*, Bristol: Intellect.

Guiette, A., Jacobs, S., Schramme, A. and Vandenbrempt, K. (2011), *Creatieve industrieën in Vlaanderen: mapping en bedrijfseconomische analyse* (Onderzoeksrapport), Antwerp: Flanders DC & Antwerp Management School.

Ha, L. and Chan-Olmsted, S. M. (2004), 'Cross-media use in electronic media: the role of cable television web sites in cable television network branding and viewership', *Journal of Broadcasting and Electronic Media*, Vol. 48, No. 4: 620–45.

Habermas, J. ([1962] 1989), *The Structural Transformation of the Public Sphere: An Inquiry into a Category of Bourgeois Society*. Cambridge, MA: The MIT Press.

Hackett, E. J., Amsterdamska, O., Lynch, M. and Wajcman, J. (eds) (2008), *The Handbook of Science and Technology Studies* (3rd edn). Cambridge, MA: The MIT Press.

Haddon, L. (1992), 'Explaining ICT consumption: The case of the home computer', in R. Silverstone and E. Hirsch (eds) *Consuming Technologies: Media and Information in Domestic Spaces*, London: Routledge, 82–96.

—(2003), 'Domestication and mobile technology', in J. E. Katz (ed.) *Machines that Become Us: The Social Context of Personal Communication Technology*, New Brunswick: Transaction Publishers, 43–56.

—(2004), *Information and Communication Technologies in Everyday Life: A Concise Introduction and Research Guide*, Oxford: Berg.

—(2006), 'The contribution of domestication research to in-home computing and media consumption', *The Information Society*, Vol. 22, No. 4: 195–204.

Hall, S. (1973), *Encoding and Decoding in the Television Discourse (Stencilled Occasional Papers)*, Birmingham: University of Birmingham.

Hamill, L. (2003), 'Time as a Rare Commodity in Home Life', in R. Harper (ed.) *Inside the Smart Home*, London: Springer, 63–78.

Hand, M., Shove, E. and Southerton, D. (2005), 'Explaining showering: a discussion of the material, conventional, and temporal dimensions of practice', *Sociological Research Online*, Vol. 10, No. 2. Available from http://www.socresonline.org.uk/10/2/hand.html

Harper, R. (ed.) (2003), *Inside the Smart Home*, London: Springer.

Hartley, J. (1999), *The Uses of Television*, London: Routledge.

Hassoun, D. (2014), 'Tracing attentions: toward an analysis of simultaneous media use', *Television and New Media*, Vol. 15, No. 4: 271–88.

Hayward, M. (2013), 'Convergence thinking, information theory and labour in "End of Television" studies', in M. de Valck and J. Teurlings (eds) *After the Break: Television Theory Today*, Amsterdam: Amsterdam University Press, 117–30.

Heeter, C., D'Alessio, D., Greenberg, B. and McVoy, S. (1988), 'Cableviewing behaviors: An electronic assessment', in C. Heeter and B. Greenberg (eds) *Cableviewing*, Norwood, NJ: Ablex Publishing Company, 51–66.

Helle-Valle, J. and Stø, E. (2003), 'Digital TV and the moral economy of the home', in M. Tarkka (ed.) *Digital Television and the Consumer Perspective* (Report from the seminar 'Digital television as consumer platform'), Faroe Islands: Torshavn, 47–54.

Hendy, D. (2000), *Radio in the Global Age*, Cambridge: Polity Press.

Hermes, J. (2013), 'Caught: Critical Versus Everyday Perspectives on Television', in M. de Valck, and J. Teurlings (eds) *After the Break: Television Theory Today*, Amsterdam: Amsterdam University Press, 35–50.

Hesmondhalgh, D. (ed.) (2006), *Media Production*, Maidenhead: Open University Press and McGraw-Hill.

—(2010), 'Media Industry Studies, Media Production Studies', in J. Curran (ed.) *Media and Society* (5th edn), London: Arnold, 57–73.

Hesmondhalgh, D. and Baker, S. (2008), 'Creative work and emotional labour in the television industry', *Theory, Culture and Society*, Vol. 25, No. 7–8: 97–118.

—(2011), *Creative Labour: Media Work in Three Cultural Industries*, Abingdon: Routledge.

Hitchens, L. (2006), *Broadcasting Pluralism and Diversity: A Comparative Study of Policy and Regulation*, Oxford: Hart Publishing.

Hobson, D. (1989), 'Soap operas at work', in E. Seiter (ed.) *Remote Control: Television, Audiences, and Cultural Power*, London: Routledge, 150–67.

Holmes, S. (2004), '"All You've got to Worry About is the Task, Having a Cup of Tea, and Doing a bit of Sunbathing": Approaching celebrity in *Big Brother*', in S. Holmes and D. Jermyn (eds) *Understanding Reality Television*, London: Routledge, 111–35.

Horowitz Associates (2014), *State of Cable and Digital Media Report* (April), New York: Horowitz Associates.

Horrocks, R. (2004), 'Turbulent television: the New Zealand experiment', *Television and New Media*, Vol. 5, No. 1: 55–68.

Hoskins, C., McFadyen, S. and Finn, A. (2004), *Media Economics: Applying Economics to New and Traditional Media*, Thousands Oaks, CA: Sage.

Hoynes, W. (2003), 'Branding public service: the "new PBS" and the privatization of public television', *Television and New Media*, Vol. 4, No. 2: 117–30. Available from http://mavise.obs.coe.int – Database on TV and on-demand audiovisual services and companies in Europe.

Huang, E., Davison, K., Shreve, S., Davis, T., Bettendorf, E. and Nair, A. (2006), 'Facing the challenges of convergence: media professionals' concerns of working across media platforms', *Convergence: The International Journal of Research into New Media Technologies*, Vol. 12, No. 1: 83–98.

Intel (2005), *Digital Lifestyle Report*, Swindon: Intel.

Iosifidis, P. (2006), 'Digital switchover in Europe', *The International Communication Gazette*, Vol. 68, No. 3: 249–68.

—(2010), *Reinventing Public Service Communication*, Basingstoke: Palgrave Macmillan.

IP Network (2008), *Television 2008: International Key Facts*, Germany: IP Network-RTL Group.

Jacobson, D. (2009), 'COPE: Create Once, Publish Everywhere'. Available from http://blog.programmableweb.com/2009/10/13/cope-create-once-publish-everywhere/

Jaffe, J. (2005), *Life After the 30-Second Spot: Energize Your Brand with a Bold Mix of Alternatives to Traditional Advertising*, Hoboken: John Wiley & Sons, Inc.

Jauss, H. R. (1982), *Toward an Aesthetic of Reception* (trans. T. Bahti), Minneapolis: University of Minnesota Press.

Jenkins, H. (2003), 'Quentin Tarantino's Star Wars? Digital cinema, media convergence, and participatory culture', in D. Thorburn H. and Jenkins (eds) *Rethinking Media Change: The Aesthetics of Transition*, Cambridge, MA: The MIT Press, 281–312.

—(2006), *Convergence Culture: Where Old and New Media Collide*, New York: New York University Press.

Jenkins, H., Ford, S. and Green, J. (2013), *Spreadable Media: Creating Value and Meaning in a Networked Culture*, New York: New York University Press.

Jensen, K. B. (2014), 'Audiences, audiences everywhere – measured, interpreted and imagined', in G. Patriarche, H. Bilandzic, J. L. Jensen and J. Jurišić (eds) *Audience Research Methodologies: Between Innovation and Consolidation*, New York: Routledge, 227–40.

Jerslev, A. (2010), 'X Factor viewers: Debate on an internet forum', in J. Gripsrud (ed.) *Relocating Television: Television in the Digital Context*, Abingdon: Routledge, 169–82.

Jin, H. (2011), 'British cultural studies, active audiences and the status of cultural theory: an interview with David Morley', *Theory, Culture and Society*, Vol. 28, No. 4: 124–44.

Johnson, C. (2012), *Branding Television*, Abingdon: Routledge.

Johnson, C. and Jones, K. (1978), *Modern Radio Station Practices*, Belmont: Wadsworth.

Johnson, V. E. (2009), 'Everything new is old again: Sport television, innovation, and tradition for a multi-platform era', in A. D. Lotz (ed.) *Beyond Prime Time: Television Programming in the Post-Network Era*, New York: Routledge, 114–37.

Jones, S. and Fox, S. (2009), *Pew Survey – Generations Online in 2009*, Washington: Pew Internet & American Life Project.

Juhlin, O., Zoric, G., Engström, A. and Reponen, E. (2014), 'Video interaction: a research agenda', *Personal and Ubiquitous Computing*, Vol. 18, No. 3: 685–92.

Kackman, M., Binfield, M., Payne, M. T., Perlman, A. and Sebok, B. (eds) (2011), *Flow TV: Television in the Age of Media Convergence*, New York: Routledge.

Kapferer, J.-N. (2012), *The New Strategic Brand Management: Advanced Insights and Strategic Thinking*, London: Kogan Page.

Katz, E. (2009), 'The end of television?', *The ANNALS of the American Academy of Political and Social Science*, Vol. 625, No. 1: 6–18.

Katz, E., Blumler, J. G. and Gurevitch, M. (1974), 'Utilization of mass communication by the individual', in J. G. Blumler and E. Katz (eds) *The Uses of Mass Communications: Current Perspectives on Gratifications Research*, Beverly Hills: Sage, 19–34.

Kavoori, A. P. (2011), *Reading YouTube: The Critical Viewers Guide*, New York: Peter Lang.

Kelley, T. and Littman, J. (2001), *The Art of Innovation: Lessons in Creativity from IDEO, America's Leading Design Firm* (1st edn), New York: Currency & Doubleday.

Kim, J. (2012), 'The institutionalization of YouTube: From user-generated content to professionally generated content', *Media, Culture and Society*, Vol. 34, No. 1: 53–67.

Kindem, G. and Musburger, R. B. (2005), *Introduction to Media Production: The Path to Digital Media Production*, Burlington: Focal Press.

Klein, B. (2009), *As Heard on TV: Popular Music in Advertising*, Farnham: Ashgate Publishing.

Kleinsteuber, H. J. (2011), *Digital Radio Broadcast (DAB): A Lost Cause in Germany*, Paper presented at 'International Association of Media and Communication Research', Istanbul, 13–17 July.

Kompare, D. (2006), 'Publishing flow: DVD box sets and the reconception of television', *Television and New Media*, Vol. 7, No. 4: 335–60.

—(2009), 'Extraordinarily ordinary: *The Osbournes* as "An American Family"', in S. Murray and L. Ouellette (eds) *Reality TV: Remaking Television Culture*, New York: New York University Press, 100–22.

—(2010), 'Reruns 2.0: revising repetition for multiplatform television distribution', *Journal of Popular Film and Television*, Vol. 38, No. 2: 79–84.

Kruse, H. (2009), 'Betting on News Corporation: interactive media, gambling, and global information flows', *Television and New Media*, Vol. 10, No. 2: 179–94.

Küng, L., Kröl, A.-M., Ripken, B. and Walker, M. (1999), 'Impact of the digital revolution on the media and communication industries', *The Public Javnost*, Vol. 6, No. 3: 29–48.

Küng, L., Picard, R. G. and Towse, R. (eds) (2008), *The Internet and the Mass Media*, London: Sage.

LaRose, R., Strover, S., Gregg, J. L. and Straubhaar, J. (2011), 'The impact of rural broadband development: lessons from a natural field experiment', *Government Information Quarterly*, Vol. 28, No. 1: 91–100.

Lash, S. and Lury, C. (2007), *Global Culture Industry: The Mediation of Things*, Malden: Polity.

Leadbeater, C. and Miller, P. (2004), *The Pro-Am Revolution: How Enthusiasts Are Changing Our Economy and Society*, London: Demos.

Lee, B. and Lee, R. S. (1995), 'How and why people watch TV: implications for the future of interactive television', *Journal of Advertising Research*, Vol. 35, No. 6: 9–18.

Lekakos, G., Chorianopoulos, K. and Doukidis, G. (2007), *Interactive Digital Television: Technologies and Applications*, Hershey, PA: IGI Publishing.

Leurdijk, A., Leendertse, M., de Munck, S., Staal, M., Vetjens, B. and de Vos, C. (2006), *Reclame 2.0: De toekomst van reclame in een digitaal televisielandschap*, Delft: TNO.

Lievrouw, L. A. and Livingstone, S. (eds) (2002), *The Handbook of New Media: Social Shaping and Consequences of ICTs*, London: Sage.

Lis, B. and Post, M. (2013), 'What's on TV? The impact of brand image and celebrity credibility on television consumption from an ingredient branding perspective', *International Journal of Media Management*, Vol. 15, No. 4: 229–44.

Lister, M., Dovey, J., Giddings, S., Grant, I. and Kelly, K. (2003), *New Media: A Critical Introduction*, London: Routledge.

Livingstone, S. (2007), 'From family television to bedroom culture: Young people's media at home', in E. Devereux (ed.) *Media Studies: Key Issues and Debates*, London: Sage, 302–21.

—(2009), 'Half a century of television in the lives of our children', *The ANNALS of the American Academy of Political and Social Science*, Vol. 625, No. 1: 151–63.

Lotz, A. D. (2007), *The Television will be Revolutionized*, New York: New York University Press.

—(2009), 'What is U.S. television now?', *The ANNALS of the American Academy of Political and Social Science*, Vol. 625, No. 1: 49–59.

Lozano, J. F. G. (2013), 'Television memory after the end of television history?', in M. de Valck and J. Teurlings (eds) *After the Break: Television Theory Today*, Amsterdam: Amsterdam University Press, 131–44.

Lull, J. (1990), *Inside Family Viewing: Ethnographic Research on Television's Audience*, London: Routledge.

Mackay, H. and O'Sullivan, T. (eds) (1999), *The Media Reader: Continuity and Transformation*, London: Sage.

MacKenzie, D. A. and Wajcman, J. (1999), *The Social Shaping of Technology* (2nd edn), Maidenhead: Open University Press.

Malmelin, N. and Moisander, J. (2014), 'Brands and branding in media management – toward a research agenda', *International Journal of Media Management*, Vol. 16, No. 1: 9–25.

Manning, S. and Sydow, J. (2007), 'Transforming creative potential in project networks: how TV movies are produced under network-based control', *Critical Sociology*, Vol. 33, No. 1: 19–42.

Mante-Meijer, E., Pierson, J. and Loos, E. (2011), 'Conclusion: Substantiating user empowerment', in J. Pierson, E. Loos and E. Mante-Meijer (eds) *New Media Technologies and User Empowerment*, Frankfurt am Main: Peter Lang, 285–307.

Marshall, P. D. (2009), 'Screens: Television's dispersed "Broadcast"', in G. Turner and J. Tay (eds) *Television Studies After TV: Understanding Television in the Post-broadcast Era*, London: Routledge, 41–50.

Matelski, M. J. (1995), 'Resilient radio', in E. C. Pease and E. E. Dennis (eds) *Radio – The Forgotten Medium*, New Brunswick: Transaction Publishers, 5–14.

McCabe, J. and Akass, K. (2008), 'It's not TV, it's HBO's original programming: Producing quality TV', in M. Leverette, B. L. Ott and C. L. Buckley (eds) *It's Not TV: Watching HBO in the Post-Television Era*, New York: Routledge, 83–94.

McCarthy, A. (2000), 'The misuse value of the TV set: reading media objects in transnational urban spaces', *International Journal of Cultural Studies*, Vol. 3, No. 3: 307–30.

McChesney, R. W. (1993), *Telecommunications, Mass Media, and Democracy: The Battle for the Control of U.S. Broadcasting, 1928–1935*. New York: Oxford University Press.

McIntosh, S. (2008), 'Will Yingshuiji buzz help HBO Asia?', in M. Leverette, B. L. Ott and C. L. Buckley (eds) *It's Not TV: Watching HBO in the Post-Television Era*, New York: Routledge, 65–82.

McLeish, R. (2005), *Radio Production*, Oxford: Focal Press.

McLuhan, M. ([1964] 1994), *Understanding Media: The Extensions of Man*, Cambridge, MA: The MIT Press.

McLuhan, M. and McLuhan, E. (1988), *Laws of Media: The New Science*, Toronto: University of Toronto Press.

McQuail, D. (1997), *Audience Analysis*, Thousand Oaks: Sage.

Meech, P. (2001), 'Corporate trails: relationship building and the BBC', *Journal of Communication Management*, Vol. 6, No. 2: 188–93.

Merton, R. K. and Lazarsfeld, P. F. (1968), 'Studies in radio and film propaganda', in R. K. Merton (ed.) *Social Theory and Social Structure*, New York: Free Press, 563–82.

Merrin, W. (2008), 'Media Studies 2.0 – My Thoughts'. Available from http://twopointzero-forum.blogspot.be/ (4 January 2008).

Meyrowitz, J. (1985), *No Sense of Place: The Impact of Electronic Media on Social Behavior*, New York: Oxford University Press.

Michalis, M. (2007), *Governing European Communications: From Unification to Coordination* (Vol. 1), Lanham: Lexington Books.

Miller, N. (2007), 'Manifesto for a new age', *Wired*, Vol. 15, No. 3. Available from http://www.wired.com/wired/archive/15.03/snackminifesto.html

Mills, P. (1985), 'An international audience?', *Media, Culture and Society*, Vol. 7, No. 4: 487–501.

Moe, H. (2010), 'Governing public service broadcasting: "public value tests" in different national contexts', *Communication, Culture and Critique*, Vol. 3, No. 2: 207–23.

Moores, S. (1993), *Interpreting Audiences: The Ethnography of Media Consumption*, London: Sage.

—(1995), 'TV discourse and "time-space distanciation" on mediated interaction in modern society', *Time and Society*, Vol. 4, No. 3: 329–44.

—(2005), *Media/Theory: Thinking about Media and Communications*, Abingdon: Routledge.

—(2012), *Media, Place, and Mobility*, Hampshire: Palgrave Macmillan.

Moran, A. (2005), 'Configurations of the New Television Landscape', in J. Wasko (ed.) *A Companion to Television*. Oxford: Wiley-Blackwell, 291–307.

Morley, D. (1986), *Family Television: Cultural Power and Domestic Leisure*, London: Comedia.

Morris, P. F. and Peterson, P. E. (2000), *The New American Democracy*, New York: Addison Wesley.

Mosco, V. (2009), *The Political Economy of Communication* (2nd edn), Los Angeles: Sage.

Mu, M., Simpson, S., Bojko, C., Broadbent, M., Brown, J., Mauthe, A., Race, N. and Hutchison, D. (2013), 'Storisphere: from TV watching to community story telling', *Communications Magazine*, Vol. 51, No. 8: 112–19.

Mullan, B. (1997), *Consuming Television: Television and its Audience*, Oxford: Blackwell Publishers.

Mullen, M. (2008), *Television in the Multichannel Age: A Brief History of Cable Television*, Malden: Blackwell Publishing.

Murdock, G. (1993), 'Communications and the constitution of modernity', *Media, Culture and Society*, Vol. 15, No. 4: 521–39.

—(2004), *Building the Digital Commons: Public Broadcasting in the Age of the Internet*. The '2004 Spry Memorial' Lecture, 22 November, University of Montreal, Montreal.

Musiani, F. (2010), *Privacy as Anonymity and Knowledge of Identity: The Dawn of Peer-to-peer Social Networks*, Paper presented at 'International Association for Media and Communication Research 2010 Annual Conference – Communication Policy and Technology Section', Braga, 18–22 July.

Napoli, P. M. (2010a), 'Revisiting "mass communication" and the "work" of the audience in the new media environment', *Media, Culture and Society*, Vol. 32, No. 3: 505–16.

—(2010b), *Audience Evolution: New Technologies and the Transformation of Media Audiences*, New York: Columbia University Press.

Newman, N. and Levy, D. A. L. (2014), *Reuters Institute Digital News Report 2014: Tracking the Future of News*, Oxford: Reuters Institute for the Study of Journalism.

Nielsen (2012), *Australian Multi-screen Report, Quarter 2. Trends in Video Viewership Beyond Conventional Television Sets*. OzTAM, Regional Television Audience Measurement (TAM) & Nielsen. Available from http://www.oztam.com.au/documents/Other/Australian%20Multi-Screen%20Report%20Q2%202013%20FINAL%20281013.pdf

Nightingale, V. (2011a), 'Introduction', in V. Nightingale (ed.) *The Handbook of Media Audiences*, Malden: Wiley-Blackwell, 1–15.

—(ed.) (2011b), *The Handbook of Media Audiences*, Malden: Wiley-Blackwell.

Nightingale, V. and Dwyer, T. (2006), 'The audience politics of "enhanced" television formats', *International Journal of Media and Cultural Politics*, Vol. 2, No. 1: 25–42.

Nightingale, V. and Ross, K. (2003), *Critical Readings: Media and Audiences*, Maidenhead: Open University Press.

Norman, D. A. (1988), *The Psychology of Everyday Things*, New York: Basic Books.

Normann, R. and Ramirez, R. (1993), 'From value chain to value constellation: designing interactive strategy', *Harvard Business Review*, Vol. 71, No. 4: 65–77.

Nyre, L. (2008), *Sound Media: from Live Journalism to Musical Recording*, London: Routledge.

O'Neill, B. and Shaw, H. (2010), 'Radio broadcasting in Europe: The search for a common digital future', in B. O'Neill, M. Ala-Fossi, P. Jauert, S. Lax, L. Nyre and H. Shaw (eds) *Digital Radio in Europe: Technologies, Industries and Cultures*, Bristol: Intellect, 27–42.

O'Neill, B., Ala-Fossi, M., Jauert, P., Lax, S., Nyre, L. and Shaw, H. (eds) (2010), *Digital Radio in Europe: Technologies, Industries and Cultures*, Bristol: Intellect.

Ofcom (2010), *The Communications Market: Digital Radio Report*, London: Ofcom. Available from http://stakeholders.ofcom.org.uk/binaries/research/radio-research/digital-radio-reports/report210710.pdf

—(2013a), *The Communications Market: Digital Radio Report* (Ofcom's fourth annual digital progress report), London: Ofcom.

—(2013b), *International Communications Market Report*, London: Ofcom.

Ollivier, H. and Pouillot, D. (eds) (2013), *DigiWorld Yearbook 2013: The Challenges of the Digital World*, Montpellier: IDATE DigiWorld Institute.

—(2014), *DigiWorld Yearbook 2014: The Challenges of the Digital World*, Montpellier: IDATE DigiWorld Institute.

Olofsson, J. (2014), 'Revisiting the TV object: on the site-specific location and objecthood of the Swedish television during its inception', *Television and New Media*, Vol. 15, No. 4: 371–86.

Örnebring, H. (2007a), 'Alternate reality gaming and convergence culture: the case of *Alias*', *International Journal of Cultural Studies*, Vol. 10, No. 4: 445–62.

—(2007b), 'The Show Must Go On … And On: Narrative and Seriality in *Alias*', in S. Abbott and S. Brown (eds) *Investigating Alias: Secrets and Spies*, London: I. B. Tauris, 11–26.

Oudshoorn, N. and Pinch, T. J. (2003), *How Users Matter: The Co-construction of Users and Technologies*, Cambridge, MA: The MIT Press.

Ouellette, L. and Murray, S. (2009), 'Introduction', in S. Murray and L. Ouellette (eds) *Reality TV: Remaking Television Culture*, New York: New York University Press, 1–22.

Owens, J. and Millerson, G. (2009), *Television Production*, Burlington: Focal Press.

Oxera (2003), *Study on Interoperability, Service Diversity and Business Models in Digital Broadcasting Markets* (Volume I: Report), Oxford: European Commission.

Papathanassopoulos, S. (2002), *European Television in the Digital Age: Issues, Dynamics and Realities*, Cambridge: Polity.

Paterson, R., Petrie, D. and Willis, J. (1995), 'Introduction', in D. Petrie and J. Willis (eds) *Television and the Household: Reports from the BFI's Audience Tracking Study*, London: British Film Institute, 1–7.

Pauwels, C. (1995), *Cultuur en economie: de spanningsvelden van het communautair audio-visueel beleid. Een onderzoek naar de grenzen en mogelijkheden van een kwalitatief cultuur- en communicatiebeleid in een economisch geïntegreerd Europa. Een kritische analyse en prospectieve evaluatie aan de hand van het gevoerde Europees audiovisueel beleid*, PhD Thesis, Vrije Universiteit Brussel, Brussels, Belgium.

—(2011), *Looking back to look forward: the end of the media world as we know it?* Presentation at 'Digital Agenda: First meeting of EU Media Futures Forum', organized by The Media Task force – EU Digital Agenda commissioner Neelie Kroes, Brussels, 7 December.

Pauwels, C. and Bauwens, J. (2007), 'Power to the people? A look at the powerlessness, dissatisfaction and lack of freedom of TV viewers', *International Journal of Media and Cultural Politics*, Vol. 3, No. 2: 149–65.

Peirce, L. M. and Tang, T. (2012), 'Refashioning television: business opportunities and challenges of webisodes', *International Journal of Business and Social Science*, Vol. 3, No. 13: 163–71.

Percival, M. J. (2011), 'Music radio and the record industry: songs, sounds, and power', *Popular Music and Society*, Vol. 34, No. 4: 455–73.

Peters, B. (2009), 'And lead us not into thinking the new is new: a bibliographic case for new media history', *New Media and Society*, Vol. 11, No. 1–2: 13–30.

Peters, S. (2003), 'Emotional Context and "Significancies" of Media', in R. Harper (ed.) *Inside the Smart Home*, London: Springer, 79–97.

Pfeiffer, M. and Zinnbauer, M. (2010), 'Can old media enhance new media? How traditional advertising pays off for an online social network', *Journal of Advertising Research*, Vol. 50, No. 1: 42–50.

Picard, R. (2011), *The Economics and Financing of Media Companies* (2nd edn), New York: Fordham University Press.

Picone, I. (2010), *Iedereen journalist? Het raadplegen, delen, beoordelen en beargumenteren van gebruikersgegenereerd nieuws – Een digitale experimentele etnografie naar evoluerende praktijken inzake nieuws bij Vlaamse nieuwsgebruikers*, PhD thesis, Vrije Universiteit Brussel, Brussels, Belgium.

Pierson, J. (2003), *De (on)verenigbaarheid van informatie- en communicatietechnologie en zelfstandige ondernemers: Een gebruikersgericht en innovatiestrategisch onderzoek naar adoptie, gebruik en betekenis van ICT voor zaakvoerders van micro-ondernemingen*, PhD thesis, Vrije Universiteit Brussel, Brussels, Belgium.

Pierson, J., Mante-Meijer, E. and Loos, E. (eds) (2011), *New Media Technologies and User Empowerment*, Frankfurt am Main: Peter Lang.

Pierson, J., Mante-Meijer, E., Loos, E. and Sapio, B. (eds) (2008), *Innovating for and by users*, Brussels: COST 298 – OPOCE.

Plunkett, J. (2009), 'Digital radio listening falls', *The Guardian* (29 January).

Pool, I. de Sola and Noam, E. M. (1990), *Technologies without Boundaries: On Telecommunications in a Global Age*, Cambridge: Harvard University Press.

Porter, M. E. (1980), *Competitive Strategy: Techniques for Analyzing Industries and Competitors*, New York: Free Press.

—(1985), *Competitive Advantage: Creating and Sustaining Superior Performance*, New York: Free Press.

Poster, M. (2005), 'Who controls digital culture?', *Fast Capitalism*, Issue 1.2. Available from http://www.fastcapitalism.com/

Putnam, R. D. (2000), *Bowling Alone: The Collapse and Revival of American Community*, New York: Simon & Schuster.

Reckwitz, A. (2002), 'Toward a theory of social practices: a development in culturalist theorizing', *European Journal of Social Theory*, Vol. 5, No. 2: 243–63.

Regal, B. (2005), *Radio: The Life Story of a Technology*, Westport: Greenwood Press.

Rideout, V. J., Foehr, U. G. and Roberts, D. F. (2010), *Generation M²: Media in the Lives of 8- to 18-Year-Olds* (A Kaiser Family Foundation Study), Menlo Park: Henry J. Kaiser Family Foundation.

Righart, H. (1995), *De eindeloze jaren zestig. Geschiedenis van een generatieconflict*, Amsterdam/Antwerpen: Uitgeverij De Arbeiderspers.

Rizzo, T. (2007), 'Programming your own channel: An archæology of the playlist', in A. T. Keynon (ed.) *TV Futures: Digital Television Policy in Australia*, Carlton: Melbourne University Press, 108–34.

Roberts, D. F. and Foehr, U. G. (2008), 'Trends in media use', *Future of Children*, Vol. 18, No. 1: 11–37.

Rogers, E. M. (2003), *Diffusion of Innovations* (5th edn), New York: Free Press.

Rose, R. L. and Wood, S. L. (2005), 'Paradox and the consumption of authenticity through reality television', *Journal of Consumer Research*, Vol. 32, No. 2: 284–96.

Rosen, C. (2004), 'The age of egocasting', *The New Atlantis: A Journal of Technology and Society*, No. 7 (Fall): 51–72.

Rosen, J. (2006), 'The people formerly known as the audience'. Available from http://journalism.nyu.edu/pubzone/weblogs/pressthink/2006/06/27/ppl_frmr.html

Rosenstein, A. W. and Grant, A. E. (1997), 'Reconceptualizing the role of habit: a new model of television audience activity', *Journal of Broadcasting and Electronic Media*, Vol. 41, No. 3: 324–44.

Scannell, P. (1989), 'Public service broadcasting and modern public life', *Media, Culture and Society*, Vol. 11, No. 2: 135–66.

—(2000), 'For-anyone-as-someone structures', *Media, Culture and Society*, Vol. 22, No. 1: 5–24.

—(2002), '*Big Brother* as a television event', *Television and New Media*, Vol. 3, No. 3: 271–82.

—(2007), *Media and Communication*, London: Sage.

—(2009), 'The dialectic of time and television', *The ANNALS of the American Academy of Political and Social Science,* Vol. 625, No. 1: 219–35.

Schatz, R., Wagner, S., Egger, S. and Jordan, N. (2007), *Mobile TV becomes social: integrating content with communications*, Paper presented at 29th International Conference on Information Technology Interfaces 'ITI 2007', Cavtat/Dubrovnik, 25–28 June.

Schoemaker, P. J. and Reese, S. D. (2014), *Mediating the Message in the 21st Century: A Media Sociology Perspective* (3rd edn), New York: Routledge.

Schreier, M. (2004), '"Please help me; all I want to know is: is it real or not?": how recipients view the reality status of *The Blair Witch Project*', *Poetics Today*, Vol. 25, No. 2: 305–34.

Schrøder, K. C. and Larsen, B. S. (2010), 'The shifting cross-media news landscape: challenges for news producers', *Journalism Studies*, Vol. 11, No. 4: 524–34.

Schwartz, E. (2003), *The Last Lone Inventor: A Tale of Deceit, Genius and the Birth of the Television*, New York: Harper.

Scolari, C. A. (2009), 'Transmedia storytelling: implicit consumers, narrative worlds, and branding in contemporary media production', *International Journal of Communication*, Vol. 3: 586–606.

Sconce, J. (2006), 'What if? Charting television's new textual boundaries', in L. Spigel and J. Olsson (eds) *Television after TV: Essays on a Medium in Transition*, Durham: Duke University Press, 93–112.

Seabright, P. and von Hagen, J. (2007), *The Economic Regulation of Broadcasting Markets: Evolving Technology and the Challenges for Policy*, Cambridge: Cambridge University Press.

Seiter, E. (2001), *Television and New Media Audiences*, Oxford: Clarendon Press.

Shao, G. (2009), 'Understanding the appeal of user-generated media: a uses and gratification perspective', *Internet Research*, Vol. 19, No. 1: 7–25.

Shaw, C. (1999), *Deciding What We Watch: Taste, Decency, and Media Ethics in the UK and the USA*, Oxford: Oxford University Press.

Silverstone, R. (1988), 'Letters of Marshall McLuhan, book review', *Media, Culture and Society*, Vol. 10, No. 3: 388–92.

—(1999), *Why Study the Media?*, London: Routledge.

—(1994a), 'Domesticating the Revolution: Information and Communication Technologies and Everyday Life', in R. Mansell (ed.) *Management of Information and Communication Technologies: Emerging Patterns of Control*, London: Aslib, 221–33.

—(1994b), *Television and Everyday Life*, London: Routledge.

Silverstone, R. and Haddon, L. (1996), 'Design and Domestication of Information and Communication Technologies: Technical Change and Everyday Life', in R. Mansell and R. Silverstone (eds) *Communication by Design: The Politics of Information and Communication Technologies*, Oxford: Oxford University Press, 44–74.

Silverstone, R. and Hirsch, E. (eds) (1992), *Consuming Technologies: Media and Information in Domestic Spaces*, London: Routledge.

Simons, N. (2013), 'Watching TV fiction in the age of digitization: a study into the viewing practices of engaged TV fiction viewers', *International Journal of Digital Television*, Vol. 4, No. 2: 177–91.

—(2014), 'Audience reception of cross- and transmedia TV drama in the age of convergence', *International Journal of Communication*, Vol. 8: 2220–39.

Smulyan, S. (1994), *Selling Radio: The Commercialization of American Broadcasting, 1920–1934*, Washington, DC: Smithsonian Publications.

Smythe, D. W. (1977), 'Communications: blindspot of Western Marxism', *Canadian Journal of Political and Social Theory*, Vol. 1, No. 3: 1–27.

—(2006), 'On the Audience commodity and its Work', in Durham, M. G. and D. M. Kellner, (eds) *Media and Cultural Studies Key Works*, Malden, MA: Blackwell, 230–56.

Snickars, P. and Vonderau, P. (eds) (2009), *The YouTube Reader*, Stockholm: National Library of Sweden & Wallflower Press.

Södergård, C. (ed.) (2003), *Mobile Television – Technology and Experiences: Report on the Mobile TV Project* (VTT Publications 506). Helsinki: VTT Technical Research Centre of Finland. Available from http://www.vtt.fi/inf/pdf/publications/2003/P506.pdf

Sonwalkar, P. (2008), 'Television in India: Growth amid a regulatory vacuum', in D. Ward (ed.) *Television and Public Policy: Change and Continuity in an Era of Global Liberalization*, New York: Lawrence Erlbaum, 115–30.

Sørensen, I. E. (2014), 'Channels as content curators: multiplatform strategies for documentary film and factual content in British public service broadcasting', *European Journal of Communication*, Vol. 29, No. 1: 34–49.

Sourbati, M. (2004), 'Digital television, online connectivity and electronic service delivery: implications for communication policy (and research)', *Media, Culture and Society*, Vol. 26, No. 4: 585–90.

Spigel, L. (1992), *Make Room for TV: Television and the Family Ideal in Post-war America*, Chicago: University of Chicago Press.

—(2001), 'Media homes: then and now', *International Journal of Cultural Studies*, Vol. 4, No. 4: 385–411.

—(2004), 'Introduction', in L. Spigel and J. Olsson (eds) *Television after TV: Essays on a Medium in Transition*, Durham: Duke University Press, 1–34.

Spigel, L. and Olsson, J. (eds) (2004), *Television after TV: Essays on a Medium in Transition*, Durham: Duke University Press.

Starks, M. (2007), *Switching to Digital Television: UK Public Policy and the Market*, Bristol: Intellect Books.

—(2013), *The Digital Television Revolution: Origins to Outcomes*, Basingstoke: Palgrave Macmillan.

Steemers, J. (1997), 'Broadcasting is dead. Long live digital choice: perspectives from the United Kingdom and Germany', *Convergence: The International Journal of Research into New Media Technologies*, Vol. 3, No. 1: 51–71.

Straubhaar, J. D. (2007), *World Television: From Global to Local*, Thousand Oaks: Sage.

Street, S. (2006), *The A to Z of British Radio*, Lanham: Scarecrow Press, Inc.

Syvertsen, T. (2003), 'Challenges to public television in the era of convergence and commercialization', *Television and New Media*, Vol. 4, No. 2: 155–75.

Taneja, H. and Mamoria, U. (2012), 'Measuring media use across platforms: evolving audience information systems', *International Journal of Media Management*, Vol. 14, No. 2: 121–40.

Taneja, H., Webster, J. G., Malthouse, E. C. and Ksiazek, T. B. (2012), 'Media consumption across platforms: identifying user-defined repertoires', *New Media and Society*, Vol. 14, No. 6: 951–68.

Tay, J. and Turner, G. (2008), 'What is television? Comparing media systems in the post-broadcast era', *Media International Australia*, No. 126: 71–82.

—(2010), 'Not the apocalypse: television futures in the digital age', *International Journal of Digital Television*, Vol. 1, No. 1: 31–50.

Taylor, A. and Harper, R. (2003), 'Switching on to switch off', in R. Harper (ed.) *Inside the Smart Home*, London: Springer, 115–26.

Taylor, T. T. (2002), 'Music and the rise of radio in 1920s America: technological imperialism, socialization, and the transformation of intimacy', *Historical Journal of Film, Radio and Television*, Vol. 22, No. 4: 425–43.

Terranova, T. (2000), 'Free labor: producing culture for the digital economy', *Social Text*, Vol. 18, No. 2: 33–58.

Terzis, G. (ed.) (2007), *European Media Governance: National and Regional Dimensions*, Bristol: Intellect Books.

The Economist (2010), 'The race to organize television: Struggling for control', *The Economist* (15 July).

—(2011), 'Radio and the internet: Tuning in – a old medium gets its digital act together', *The Economist* (14 April)

—(2013), 'Schumpeter: The real Disney', *The Economist* (30 March).

Thierer, A. and Eskelsen, G. (2008), *Media Metrics: The True State of the Modern Media Marketplace*, Washington, DC: The Progress and Freedom Foundation.

Thompson, J. B. (1995), *The Media and Modernity: A Social Theory of the Media*, Cambridge: Polity Press.

Tichi, C. (1992), *Electronic Hearth: Creating an American Television Culture*, Oxford: Oxford University Press.

Toffler, A. (1980), *The Third Wave*, New York: Morrow.

Tunstall, J. (1971), *Journalists at Work*, London: Constable.

—(1993), *Television Producers*, London: Routledge.

Turner, G. (2006), 'The mass production of celebrity: celetoids, reality TV and the "demotic turn"', *International Journal of Cultural Studies*, Vol. 9, No. 2: 153–65.

—(2011), 'Convergence and divergence: The international experience of digital television', in J. Bennett and N. Strange (eds) *Television as Digital Media*, Durham: Duke University Press, 31–51.

Turner, G. and Tay, J. (eds) (2009), *Television Studies after TV: Understanding Television in the Post-broadcast Era*, London: Routledge.

Turpeinen, M. (2003), 'Co-Evolution of broadcast, customized and community-created media', in G. F. Lowe and T. Hujanen (eds) *Broadcasting and Convergence: New Articulations of the Public Service Remit*, Göteborg: Nordicom, 301–12.

UBA (2014), *Jaarverslag 2013*, s.l.: UBA.

Uricchio, W. (2004a), 'Storage, simultaneity, and the media technologies of modernity', in J. Olsson and J. Fullerton (eds) *Allegories of Communication: Intermedial Concerns from Cinema to the Digital*, Eastleigh: John Libbey, 123–38.

—(2004b), 'Television's next generation: Technology/interface culture/flow', in L. Spigel and J. Olsson (eds) *Television After TV*, Durham: Duke University Press, 232–61.

—(2009), 'Contextualizing the broadcast era: nation, commerce, and constraint', *The ANNALS of the American Academy of Political and Social Science*, Vol. 625, No. 1: 60–73.

—(2013), 'Constructing television: Thirty years that froze an otherwise dynamic medium', in M. de Valck and J. Teurlings (eds) *After the Break: Television Theory Today*, Amsterdam: Amsterdam University Press, 65–78.

Ursell, G. (2000), 'Television production: Issues of exploitation, commodification and subjectivity in UK television labour markets', *Media, Culture and Society*, Vol. 22, No. 6: 805–25.

—(2006), 'Working in the media', in D. Hesmondhalgh (ed.) *Media Production*, Maidenhead: Open University Press & McGraw-Hill Education, 133–72.

Valcke, P. and Stevens, D. (2007), 'Graduated regulation of "regulatable" content and the European Audiovisual Media Services Directive: one small step for the industry and one giant leap for the legislator', *Telematics and Informatics*, Vol. 24, No. 2, 285–302.

Van den Broeck, W. (2010), *From analogue to digital: the silent (r)evolution? A qualitative study on the domestication of interactive digital television in Flanders*, PhD thesis, Vrije Universiteit Brussel, Brussels, Belgium.

Van den Broeck, W., Lievens, B. and Pierson, J. (2006), *Domestication research for media and technology development: a case study*. Paper presented at 'Audience section at the International Association for Media and Communication Research conference "Knowledge Societies for All: Media and Communication Strategies"', Cairo, 23–28 July.

Van den Broeck, W., Bauwens, J. and Pierson, J. (2011), 'The promises of iDTV: between push marketing and consumer needs', *Telematics and Informatics*, Vol. 28, No. 4: 230–38.

Van den Broeck, W. and Pierson, J. (eds) (2008), *Digital Television in Europe*, Brussels: VUB Press.

—(2001), 'Public service television and national identity as a project of modernity: the example of Flemish television', *Media, Culture and Society*, Vol. 23, No. 1: 53–69.

—(2007), 'Old ideas meet new technologies: will digitalization save public service broadcasting (ideals) from commercial death?', *Sociology Compass*, Vol. 1, No. 1: 28–40.

Van den Bulck, H. (2008), 'Can PSB stake its claim in a media world of digital convergence? The case of the Flemish PSB Management contract renewal from an international perspective', *Convergence: The International Journal of Research into New Media Technologies*, Vol. 14, No. 3: 335–49.

Van den Bulck, H. and Enli, G. S. (2014), 'Bye bye "hello ladies?" in-vision announcers as continuity technique in a European postlinear television landscape: the case of Flanders and Norway', *Television and New Media*, Vol. 15, No. 5: 453–69.

Van den Dam, R. and Nelson, E. (2008), 'How telcos will change advertising', *Journal of Telecommunications Management*, Vol. 1, No. 3: 237–246.

Van den Eede, Y. (2012), *Amor Technologiae: Marshall McLuhan as Philosopher of Technology – Toward a Philosophy of Human–Media Relationships*, Brussels: VUBPress.

van Dijck, J. (2007a), *Television 2.0: YouTube and the Emergence of Homecasting*, Paper presented at 'MiT5 – Media in Transition: Creativity, ownership and collaboration in the digital age', Massachusetts Institute of Technology, Cambridge, MA, 27–29 April.

—(2007b), 'YouTube Beyond Technology and Cultural Form', in M. de Valck and J. Teurlings (eds) *After the Break: Television Theory Today*, Amsterdam: Amsterdam University Press, 147–60.

—(2008), 'Future memories: the construction of cinematic hindsight', *Theory, Culture and Society*, Vol. 25, No. 3: 71–87.

Van Thillo, C. (2011), *Broadcasting and publishing*, Presentation at PTV20'EU Conference – Private Television in Europe – 20 years of television without frontiers and beyond, IBBT-SMIT & VUB-IES, Brussels, 28 April.

Vandenbrande, K. (2002), *Verscholen achter de krant: media, nieuws en burgerschap in het dagelijks leven – Een publieksonderzoek naar de betekenis en beleving van de krant in een gemediatiseerde laat-moderne samenleving*, PhD thesis, Vrije Universiteit Brussel, Brussels, Belgium.

Vangenck, M., Jacobs, A., Lievens, B., Vanhengel, E. and Pierson, J. (2008), 'Does mobile television challenge the dimension of viewing television? An explorative research on time, place and social context of the use of mobile television content', in M. Tscheligi, R. Bernhaupt, L. van de Wijngaert, M. Obrist, E. Beck and S. Kepplinger (eds) *Changing Television Environments* (Vol. 5066/2008), Berlin: Springer, 122–7.

van Zoonen, L. and Aalberts, C. (2004), *Televisiekijken in het digitale tijdperk*, Paper presented at 'SISWO Conferentie 2004', Amsterdam, 23 April.

Voorveld, H. A. M. and van der Goot, M. (2013), 'Age differences in media multitasking: a diary study', *Journal of Broadcasting and Electronic Media*, Vol. 57, No. 3: 392–408.

Waisbord, S. (ed.) (2014), *Media Sociology: A Reappraisal*, Cambridge: Polity Press.

Walravens, N. and Pauwels, C. (2011), 'From high hopes to high deficit and back: a historic overview of Europe's HDTV policy and reflections towards the future of HDTV', *Telematics and Informatics*, Vol. 28, No. 4: 283–94.

Wasko, J. and Erickson, M. (2009), 'The political economy of YouTube', in P. Snickars and P. Vonderau (eds) *The YouTube Reader*, Stockholm: National Library of Sweden/Wallflower Press, 372–86.

Wasko, J., Murdock, G. and Sousa, H. (2011), *Handbook of Political Economy of Communications*, Chichester: Wiley-Blackwell.

Webster, J. and Ksiazek, T. B. (2012), 'The dynamics of audience fragmentation: public attention in an age of digital media', *Journal of Communication*, Vol. 62, No. 1: 39–56.

Webster, J. G. (2005), 'Beneath the veneer of fragmentation: television audience polarization in a multichannel world', *Journal of Communication*, Vol. 55, No. 2: 366–382.

Webster, J. G. and Phalen, P. F. (1997), *The Mass Audience: Rediscovering the Dominant Model*, Mahwah, NJ: Lawrence Erlbaum Associates.

Weiser, M. (1991), 'The computer for the 21st century', *Scientific American*, September, Vol. 265, No. 3: 94–104.

Westlund, O. and Bjur, J. (2014), 'Media life of the young', *Young*, Vol. 22, No. 1: 21–41.

Whitworth, B. and de Moor, A. (2009), *Handbook of Research on Socio-technical Design and Social Networking Systems (Vols I and II)*, Hershey, PA: IGI Information Science Reference.

Wicks, J. L., Sylvie, G., Hollifield, C. A., Lacy, S. and Sohn, A. (2004), *Media Management: A Casebook Approach*, Mahwah, NJ: Lawrence Earlbaum Associates.

Williams, R. ([1961] 2001), *The Long Revolution*, Peterborough: Encore Editions from Broadview Press.

—([1974] 2003), *Television: Technology and Cultural Form*, London: Routledge.

Williams, R. and Edge, D. (1996), 'The social shaping of technology', *Research Policy*, Vol. 25: 865–99.

Wilson, T. (2009), *Understanding Media Users: From Theory to Practice*, Chichester: Wiley-Blackwell.

Winseck, D. R. (2011), 'Introductory essay: The political economies of media and the transformation of the global media industries', in D. R. Winseck and D. Y. Jin (eds) *The Political Economies of Media: The Transformation of the Global Media Industries*, London: Bloomsbury Academic, 3–48.

Winseck, D. R. and Jin, D. Y. (eds) (2011), *The Political Economies of Media: The Transformation of the Global Media Industries*. London: Bloomsbury Academic.

Winston, B. (1998), *Media Technology and Society: A History – from the Telegraph to the Internet*, London: Routledge.

Wood, H. (2007), 'Television is happening: methodological considerations for capturing digital television reception', *European Journal of Cultural Studies*, Vol. 10, No. 4: 485–506.

World DAB (2005), 'New wave of DAB legislation and developments worldwide'. Available from http://www.worlddab.org/

—(2009), *Global Broadcasting Update. DAB/DAB+/DMB*, London: World DMB.

Yang, L. and Bao, H. (2012), 'Queerly intimate: friends, fans and affective communication in a Super Girl fan fiction community', *Cultural Studies*, Vol. 26, No. 6: 842–71.

Yoshimi, S. (2010), 'Japanese television: Early development and research', in J. Wasko (ed.) *A Companion to Television*, Malden: Blackwell Publishing, 540–57.

Ytreberg, E. (2009), 'Extended liveness and eventfulness in multi-platform reality formats', *New Media and Society*, Vol. 11, No. 4: 467–85.

Yuan, E. J. and Webster, J. G. (2006), 'Channel repertories: using peoplemeter data in Beijing', *Journal of Broadcasting and Electronic Media*, Vol. 50, No. 3: 524–36.

Zelizer, B. (1993), *Covering the Body: The Kennedy Assassination, the Media and the Shaping of Collective Memory*, Chicago: Chicago University Press.

Zerdick, A., Picot, A., Schrappe, K., Artopé, A., Goldhammer, K., Lange, U. T., Vierkant, E., Lopez-Escobar, E. and Silverstone, R. (2000), *E-conomics: Strategies for the Digital Marketplace*, Berlin: Springer.

Zettl, H. (2012), *Television Production Handbook*, Stanford: Cengage Learning.

INDEX

Milton Keynes UK
Ingram Content Group UK Ltd.
UKHW020852161024
449580UK00005B/58